Th[illegible] rura[illegible] and 'urban' is one of the oldest ideas in Geography and is deeply engrained in our culture. Throughout history, the rural has been attributed with many meanings: as a source of food and energy; as a pristine wilderness, or as a bucolic idyll; as a playground, or a place of escape; as a fragile space of nature, in need of protection; and as a primitive place, in need of modernization. But is the idea of the rural still relevant today?

Rural provides an advanced introduction to the study of rural places and processes in Geography and related disciplines. Drawing extensively on the latest research in rural geography, this book explores the diverse meanings that have been attached to the rural, examines how ideas of the rural have been produced and reproduced, and investigates the influence of different ideas in shaping the social and economic structure of rural localities and the everyday lives of people who live, work or play in rural areas.

This authoritative book contains case studies drawn from both the developed and the developing world to introduce and illustrate conceptual ideas and approaches, as well as suggested further reading. Written in an engaging and lively style, *Rural* challenges the reader to think differently about the rural.

Michael Woods is Professor of Human Geography in the Institute of Geography and Earth Sciences, Aberystwyth University. He specializes in rural geography, political geography and contemporary rural politics and governance.

Key Ideas in Geography

Series editors: Sarah Holloway, Loughborough University and Gill Valentine, Sheffield University

The Key Ideas in Geography series will provide strong, original, and accessible texts on important spatial concepts for academics and students working in the fields of geography, sociology and anthropology, as well as the interdisciplinary fields of urban and rural studies, development and cultural studies. Each text will locate a key idea within its traditions of thought, provide grounds for understanding its various useages and meanings, and offer critical discussion of the contribution of relevant authors and thinkers.

Published

Nature
NOEL CASTREE

City
PHIL HUBBARD

Home
ALISON BLUNT AND ROBYN DOWLING

Landscape
JOHN WYLIE

Mobility
PETER ADEY

Migration
MICHAEL SAMERS

Scale
ANDREW HEROD

Rural
MICHAEL WOODS

RURAL

Michael Woods

Routledge
Taylor & Francis Group
LONDON AND NEW YORK

First published 2011
by Routledge
2 Park Square, Milton Park, Abingdon, Oxon, OX14 4RN

Simultaneously published in the USA and Canada
by Routledge
270 Madison Avenue, New York, NY 10016

Routledge is an imprint of the Taylor & Francis Group, an informa business

Typeset in Joanna MT by Glyph International Ltd.
Printed and bound in Great Britain by CPI Antony Rowe, Chippeham, Wiltshire

British Library Cataloguing in Publication Data
A catalogue record for this book is available from the British Library

Library of Congress Cataloging-in-Publication Data
Woods, Michael, 1946–
Rural / by Michael Woods.
p. cm.
Includes bibliographical references and index.
ISBN 978-0-415-44239-8 (hbk) – ISBN 978-0-415-44240-4 (pbk)
1. Rural geography. 2. Geographical perception. I. Title.
GF127.W68 2010
307.72–dc22 2010012698

ISBN 13: 978-0-415-44239-8 (hbk)
ISBN 13: 978-0-415-44240-4 (pbk)
ISBN 13: 978-0-203-84430-4 (ebk)

In memory of Bill Edwards (1944–2007)

Contents

List of illustrations

PLATES

FIGURES

TABLES

BOXES

Acknowledgements

This book tries to capture something of the essence and excitement of contemporary rural geography, and as such it necessarily builds on the diverse work of rural geographers, sociologists and other scholars engaged in studying the rural. I have been inspired and guided in the selection of themes and topics, narratives, case studies and references, by many colleagues in the rural studies community. Where I have drawn on published work, this has been appropriately cited, but I am also grateful for the many prompts and pointers provided by conference papers and informal conversations, which cannot be easily cited here.

My approach in this book has also been shaped by the vibrant intellectual atmosphere of the Institute of Geography and Earth Sciences at Aberystwyth University, and especially by conversations with colleagues including Chris Bear, Deborah Dixon, Kate Edwards, Graham Gardner, Matt Hannah, Jesse Heley, Gareth Hoskins, Laura Jones, Martin Jones, Rhys Jones, Pete Merriman, Heidi Scott, Suzie Watkin, Mark Whitehead and Sophie Wynne-Jones. In particular, I am indebted in much of my thinking about the rural to conversations with Bill Edwards, whose avuncular counsel in British rural geography is much missed, and to whom this volume is dedicated.

I am also grateful to the School of Social Science at the University of Queensland, where a large part of this book was written during study leave, and especially to Lynda Cheshire, Geoff Lawrence and Carol Richards. At Routledge, Andrew Mould, Michael Jones and Faye Leerink are to be thanked for their patient yet persistent prompting and advice. I am further indebted to Sarah Holloway and three anonymous reviewers for their very helpful remarks on an earlier version of the manuscript.

Finally, the following publishers are thanked for permission to reproduce figures in this volume:

Figure 1.1 with kind permission of Sage Publishing.
Figure 5.1 with kind permission of Royal Van Gorcum B.V.
Figure 6.1 with kind permission of Elsevier.

1

APPROACHING THE RURAL

WHY RURAL?

Rural space has many functions and many meanings. Rural areas produce most of the world's food, and capture most of its water supply. They are the source of most of our energy – whether from fossil fuels or renewable resources – and the origin of most of the minerals that feed industry. Historically, at least, rural areas have provided society with fibre for clothing, stone and timber for building, and wood pulp to make paper. Rural areas have also become our playground – a place to walk, ride, cycle, sightsee, or simply escape in search of a slice of tranquillity. They are valued for their scenic landscapes and for their natural environments – rural areas host the vast majority of the globe's plant and animal species. Rural areas are also home to diverse indigenous cultures, and can be venerated as places where elements of traditional, pre-industrial ways of life may be glimpsed. As such, rural areas are frequently endowed with symbolic importance as signifiers of national identity, or as the counterpoint to modernity. Rural areas are celebrated variously both as wilderness and as a bucolic idyll. Yet, they can also be portrayed as remote, backward, under-developed places, in need of modernization.

The varied functions and meanings that have been attributed to rural space have made the rural into an ambiguous and complex concept. The rural is a messy and slippery idea that eludes easy definition and demarcation.

We could probably all instinctively say whether any given place was rural to us, rather than urban, but explaining why it was rural, not urban, and drawing a boundary line between urban and rural space on a map are altogether more difficult tasks. As different individuals will disagree on the meaning of rurality, and on the emphasis to be placed on different functions of rural space, so the rural is recast as a heavily contested space.

Indeed, it is the complex and contested nature of the rural that has positioned rural space as central to many key issues facing contemporary society. Debates about global food supply, for example, may be articulated through the urban-based media and political arenas, but they directly concern the management of rural space (Plate 1.1). The challenge of ensuring global food security demands that we consider the extent to which food production should be prioritized over other uses of rural land, and whether we are prepared to pursue more intensive and hi-tech forms of farming (such as genetically modified crops) that carry both environmental risks and threats to traditional social structures, such as the family farm. Similarly, pressing issues of energy security, adaptation to climate change, tackling global poverty, controlling migration, preserving biodiversity and respecting indigenous cultures all raise difficult questions about the meaning, function and management of rural space.

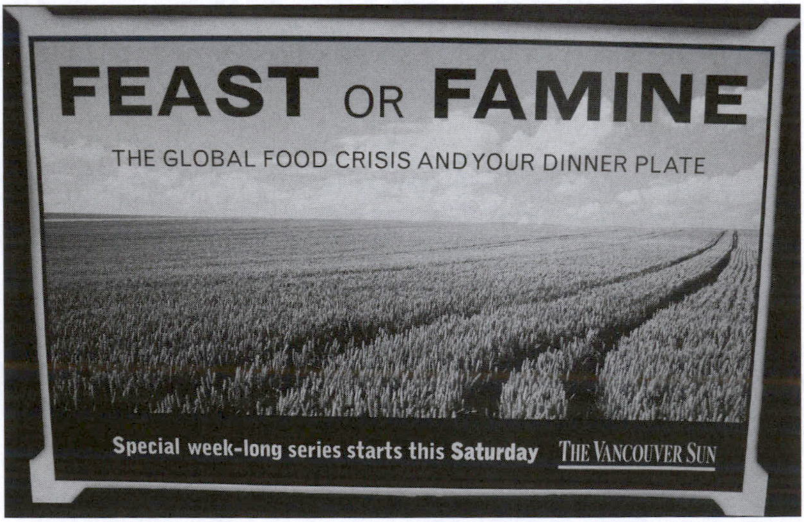

Plate 1.1 The 'global food crisis' connects urban and rural (Photo: author)

This book is about the meanings attributed to the rural, and how these diverse meanings have shaped the social and economic structure of rural localities and the everyday lives of people who live, work or play in rural areas. It discusses studies by rural geographers that have examined these processes and their effects, and also reflects on the ways in which the study of rural geography has itself been informed by different ideas about rurality. This book is intended to be forward-looking, capturing the richness and breadth of contemporary research in rural geography, but also anticipates some of the themes and approaches that will be the focus of rural geographical inquiry in coming years. In order to do this, however, it is first necessary to consider the historical development of ideas about the rural, and their application in rural geography. The remainder of this introductory chapter consequently presents a brief overview of the development of rural geography as a field of academic study and of the conceptualization of the rural within rural studies. It also considers briefly the origins of the term 'rural' and its usage in popular language – an analysis that is developed further in the next chapter, which presents a more detailed examination of the production and reproduction of the rural as an idea, its imagination and representation in popular culture, and its translation into material form in the landscape.

STUDYING THE RURAL

The city and the country

The distinction between 'urban' and 'rural', between the city and the country, is one of the oldest and most pervasive of geographical binaries. The terms may have originated as a way of differentiating between the enclosed and defensible spaces of early towns (Box 1.1), and the open and uncontrollable spaces that lay outside, but they soon acquired greater symbolic significance as they became embedded in language and culture. As Raymond Williams observed, '"country" and "city" are very powerful words, and this is not surprising when we remember how much they seem to stand for in the experience of human communities' (1973: 1). For Williams, the two terms were inextricably connected and the connection represented the progression of human society. As such, he noted, both the 'country' and the 'city', the 'rural' and the 'urban', had collected powerful feelings and associations:

Box 1.1 RURAL, RURALITY, COUNTRY AND COUNTRYSIDE

The English language uses several terms that relate to the idea of rurality or to rural space. The pairings of rural/rurality and country/countryside can be traced back to different Latin roots which both reflect something of the character of being rural. 'Rural' emerged as an adjective linked to the Latin noun *rus*, meaning an open area (Ayto, 1990). The adjective form stuck and was incorporated into several European languages to refer to something relating to those areas outside cities. The derivative 'rurality' appeared in the eighteenth century to refer to the condition of being rural, probably borrowed from the equivalent French term *ruralité*. 'Rural' was also used for a time as a noun to refer to people from rural areas, and has more recently been employed again as a noun in academic writing to refer to an abstract space that exhibits the characteristics of being rural but is not necessarily tied to a particular territory – as in the chapter headings for this book.

The word 'country' derives from the Latin preposition *contra* or 'against', and in its original Latin form originally meant 'the land spread out around one'. It hence became used to refer to an area of land, and subsequently both to the land set against the town and to the land belonging to a particular people or nation, and these two usages have remained closely associated. The term 'countryside' originally emphasized the definition of the country relative to the town (the 'side' of the town), but expanded to take on a broader, and symbolically laden, meaning in British popular culture (Bunce, 2003). However, as Bunce (2003) notes, 'countryside' does not have the same emotional charge in other English-speaking countries, where it has not been widely used, at least until recently. Where 'countryside' is used in North America, for example, it tends to retain more of its original meaning, being mainly applied to rural areas close to urban centres in regions such as New England and southern Ontario.

'Rural', 'rurality' and 'rural areas' tend to be preferred in academic and official usage over 'country' and 'countryside' probably because they sound like more neutral, objective terms, with 'country' and 'countryside' arguably more redolent with cultural meaning (interestingly, 'rural' was once synonymous with 'rustic' but the latter term has evolved a more specific meaning). There are though few rational grounds for this, and 'rural' is commonly used as the adjective for the countryside. Take, for example, the government bodies responsible for rural policy in England: the Countryside Commission and the Rural Development Commission were merged to form the Countryside Agency, which then became in reduced form the Commission for Rural Communities.

'Rural' has the advantage of being common to several European languages, including English, French, Italian and Spanish, whereas many languages do not have a direct equivalent to the term 'countryside'. Significantly, many languages employ terms to refer to rural areas, or rural people or the rural landscape that emphasize either connection to the land and agriculture, or to national identity. Thus, German uses *landschaft* (countryside or landscape) and *ländlich* (rural); French uses *paysan* (country person) and *paysage* (landscape), which are linked to *pays* (nation); and Spanish has *campestre* (rural) and *campesino* (rural person), linked to *campo* (field).

Further reading: Bunce (2003), Williams (1973).

> On the country has gathered the idea of a natural way of life: of peace, innocence and simple virtue. On the city has gathered the idea of an achieved centre: of learning, communication, light. Powerful hostile associations have also developed: on the city as a place of noise, worldliness and ambition; on the country as a place of backwardness, ignorance, limitation. A contrast between country and city, as fundamental ways of life, reaches back into classical times.
>
> (Williams, 1973: 1)

The development of 'rurality', or 'the country', as an idea in popular culture is discussed further in Chapter 2. For the moment, it is sufficient to note that the binary of 'urban' and 'rural' has also been incorporated into the organization of geography as an academic discipline, and that the popular cultural associations of the city and the country have been influential in setting the parameters of 'urban geography' and 'rural geography' and in defining their objects of inquiry.

In the early development of geography as an academic discipline, during the first two-thirds of the twentieth century, the study of both the city and the country were linked within the broader approach of 'regional geography'. This approach sought to describe the geographical characteristics of specified regions and in doing so tended to reproduce popular assumptions about the relationship between the city and the country. Thus, the geographies of rural areas tended to be described and explained in terms of their functional relationship to urban centres as sources of food and natural resources. Attempts were made to convert these popular perceptions into scientific theories by producing general models of the relationships between urban and rural areas, which could in theory be applied to any region. Such models included von Thünen's concentric model of land use, which mapped out types of farming in relation to the proximity of rural areas to cities (originally designed by a German economist Johann Heinrich von Thünen in 1826, but not translated into English until 1966); and 'central place theory' developed by Walter Christaller in 1933 (and modified by August Lösch in 1954) to explain the hierarchy of rural and urban settlements.

In practice, these models failed to capture the diversity and dynamism of rural areas, and frequently did not fit when applied empirically. Nonetheless, they prefigured the development of a new systems-based approach in geography in the 1960s that critiqued regional geography for being overly descriptive and lacking scientific rigour. Applying positivist principles of scientific inquiry and interrogating quantitative data to identify patterns and laws of spatial organization, the new 'spatial science' began to focus in on the city as its major object of research (Hubbard, 2006). The emergent 'urban geography' concerned itself with mapping and modelling 'urban systems', which could extend to and encompass rural areas, but which in effect marginalized the rural as an adjunct to the urban. Significantly, there was no equivalent investigation of 'rural systems', but rather the development of a 'systematic agricultural geography' that reinforced the association of the rural and farming (Woods, 2009a).

It was not until the early 1970s that an integrated approach to studying 'rural geography' was articulated, notably in textbooks by Clout (1972) in Britain, and Hart (1974) in the United States. These books recognized that the rural was more than agriculture, but they nonetheless presented the rural as a coherent and distinctive system, centred on productive land uses. This tension remained apparent in the new 'rural geography' that these interventions inspired. On the one hand, rural geography uncritically accepted the existence of 'rural space' as a container for the phenomena that they studied, yet, on the other hand, their attempts to distil the essence of the rural, and to authoritatively map the boundaries of rural and urban space, were compromised by methodological problems in fixing the scale of analysis, by the arbitrary spatial units of available data, and by the arbitrary nature of the indicators selected (Cloke, 2006).

Conceptual developments in rural geography

The trajectory followed by rural geography since the 1970s has been strongly influenced by wider conceptual developments in human geography (and in the social sciences more widely), in the shifting focus of its objects of study, in the explanations that it has presented for the processes and phenomena observed, and in its definition of the rural. Early research in rural geography, as described above, followed the principles of positivism, which held that objective facts could be uncovered through empirical inquiry. As such, rural geographers sought to objectively define the rural by searching for the functional characteristics that could be statistically proven to be different from urban characteristics. However, as Cloke (2006) demonstrates, the functional approach was flawed in its assumptions and undermined by its methodological weaknesses. Functional concepts of rurality could describe the characteristics of specific rural spaces and rural societies, but they could not prove that such characteristics were intrinsically rural, or explain how these characteristics shaped the realities of rural life.

The inadequacies of the functional approach were further exposed by the development of a new wave of studies in the 1970s and 1980s that adopted a political-economy approach, influenced by neo-Marxist theories of the operation of capitalism (see Buttel and Newby, 1980; Cloke, 1989a; Woods, 2005a, 2009a). Some of these studies contributed, alongside work in rural sociology, to a political-economy analysis of agriculture that emphasized the structuring of farming as a capitalist industry, subject to the same imperatives for capital accumulation as other industries. There was no place in this perspective for nostalgic and romanticized ideas about farming as the centre of a traditional rural way of life, as could still be found in positivist studies of rural geography. Other political-economy research interrogated rural planning and economic development processes as expressions of the capitalist state, examined urban to rural shifts in manufacturing and service sector employment as realignments in the capitalist spatial division of labour, and studied community social relations and the impacts of migration and population change through the prism of class analysis (Cloke, 1989a; Woods, 2009a).

These studies demonstrated that the processes shaping contemporary rural spaces and societies transcended the supposed boundaries of rural space, operating at regional, national and global scales. The impact of wider social and economic processes on particular rural localities is mediated by local factors, producing uneven development, but these local factors will vary between different rural localities, just as they will vary between urban localities. As such, the explanatory capacity of the rural–urban dualism and the value of 'rural' as a geographical concept, was brought into question:

> The broad category 'rural' is obfuscatory, whether the aim is description or theoretical evaluation, since intra-rural differences can be enormous and rural–urban similarities can be sharp.
>
> (Hoggart, 1990: 245)

The logical outcome of this critique would have been to 'do away with rural' (*Ibid*) as a meaningful concept in human geography. Yet, whatever the difficulties experienced by geographers in attempting to delimit rural space or ascribe the condition of being rural with explanatory powers, it was clear that the idea of rurality continued to be widely recognized and employed within the general population and that 'rural' continued to have a very clear and powerful meaning for many people.

A framework for exploring these meanings was provided by the 'cultural turn' in human geography and the introduction of post-modern and post-structuralist theories into rural geography. In contrast to both positivist and political-economy perspectives, post-modern theory holds that there is no objective truth waiting to be discovered. What matters is the way in which individuals, and institutions, construct their own realities in order to make sense of the world. Following this principle, rural geographers started to deconstruct the ways in which dominant ideas about rurality had been produced and reproduced (see Chapter 2), as well as exploring alternative experiences and meanings of rurality articulated by subordinate groups. In one especially influential intervention, Philo (1992) criticized the tendency of rural geography research to portray rural people as 'Mr Averages' – men in employment, white, without sexuality, healthy and able in body, and devoid of religious or political identity – and called for engagement with the 'neglected rural geographies' of other social groups beyond this stereotype.

The attention of rural geographers accordingly also began to shift away from the structural characteristics and dynamics of rural localities, to representations of the rural. In this new approach, rurality is understood as a social construct – that is as an imagined entity that is brought into being by particular discourses of rurality that are produced, reproduced and contested by academics, the media, policy-makers, rural lobby groups and ordinary individuals. The rural is therefore 'a category of thought' (Mormont, 1990: 40).

Towards a three-fold model of rural space

Social constructions of rurality reference material objects, practices and places, but they are not tied to them. As Halfacree (1993) has suggested, the proliferation of diverse representations of rural space means that the sign of the 'rural' is becoming increasingly detached from the referent of rural geographical space. In other words, the way in which the countryside is imagined in popular discourses may have little correspondence with the actual 'realities' of rural space and rural life. The world teems with virtual ruralities, ideas of the rural that are not grounded in concrete places or lived experience, and yet, such is the power and popularity of these ideas that attempts are made to bend rural space to fit their image:

> If at some time in the past some 'real' form of rurality was responsible for cultural mappings of rurality, it may now be the case that cultural mappings precede and direct the recognition of rural space, presenting us with some kind of virtual rurality.
>
> (Cloke, 2006: 22)

A framework for exploring these contingent and complex relations between representations of the rural, rural localities and the lived experiences of rural life, has been proposed by Keith Halfacree (2006), in what he describes as a strategy for bringing 'together the dispersed elements of what we already know about rural space' (p. 44). Halfacree argues that rural space is imaginative, material *and* practised, thus cutting across the polarity of locality-based and social representation-based approaches to defining rurality. He contends that these two approaches are inter-woven as 'the material and ideational rural spaces they refer to intersect in practice' (p. 47). Social representations of rurality cannot exist without imagining some form of rural locality, whilst the definition of rural localities relies upon the actualization of particular ideas about what rurality should be like. Furthermore, Halfacree notes that both material and ideational rural spaces are brought into being by practice:

> We must note how the material space of the rural locality only exists through the practices of structural processes, and how the ideational space of rural social representations only exists through the practices of discursive interaction.
>
> (Halfacree, 2006: 48)

From these initial observations, Halfacree draws on the theories of space propagated by Henri Lefebvre (which propose that space is produced and reproduced through capitalism, moulded by the pressures of the market and social reproduction, 'colonized and commodified, bought and sold, created and torn down, used and abused, speculated and fought over' (Merrifield, 2000: 173, see also Lefebvre, 1991)), to outline a 'three-fold model of rural space' (Figure 1.1). This contends that rural space comprises three intermeshed facets (Halfacree, 2006: 51):

- *Rural localities* inscribed through relatively distinctive spatial practices linked to either production or consumption.
- *Formal representations of the rural*, such as those expressed by capitalist interests or politicians, which refer to the ways in which the rural is framed within capitalist processes of production and exchange.
- *Everyday lives of the rural*, incorporating both individual and social elements in the negotiation and interpretation of rural life, and which are 'inevitably incoherent and fractured' (*Ibid*).

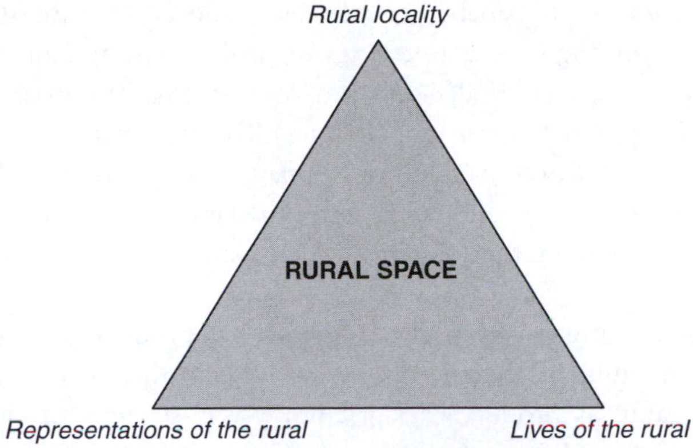

Figure 1.1 The three-fold model of rural space (after Halfacree, 2006) (by permission of Sage Publishing)

These three facets collectively make up the totality of rural space, but they do not necessarily cohere to consistently produce a congruent and unified rurality. Tensions exist between the forces of permanence and flow, as well as between the autonomous logics of the three facets. Thus, for example, 'formal representations never completely overwhelm the experience of everyday life – although they may come close – and the extent to which formal representations and local spatial practices are unified is also uneven' (Halfacree, 2006: 51–52). These tensions drive the dynamism of rural space, enabling the opportunities for rural restructuring, and creating the space for a 'politics of the rural' in which the meaning and regulation of the rural is the core issue of debate (Woods, 2003a).

Halfacree illustrates his model through the example of the centrality of productivist agriculture post-war rural Britain. As is discussed further in Chapter 3, productivism was a policy discourse that supported systems to maximize agricultural production, but became the cornerstone of British rural life from the late 1940s to the 1980s, articulated through all three facets of Halfacree's model. First, '*rural locality* was inscribed through the predominance of particular agricultural practices' (Halfacree, 2006: 53), especially industrialized forms of farming which in turn impacted on the wider social, economic and environmental elements of the locality. Second, productivism was underpinned by *formal representations of the rural* in the form of legislation and policy documents, notably the Scott Report on Land Utilization in Rural Areas in 1942, the 1947 Agriculture Act and various policy 'white papers' periodically produced over the three decades. Third, 'through the connections of productivist agriculture to the wider civil society of rural places, *everyday lives of the rural* existed largely through this productivist vision' (*Ibid*: 54). This was manifested not only in the living and working conditions of farm households, but also in the involvement of farmers and landowners in shaping the policies of rural local government in their interests (see Newby *et al.*, 1978; Woods, 2005b).

Initially, the three facets meshed together well producing a largely congruent and unified coherence in which 'the formal representation of British rurality as productivist agriculture was strongly unified, quite overwhelming and fairly hegemonic' (Halfacree, 2006: 54). Yet, over time, 'each facet of this productivist rural space *was* contested by other spaces, rural and non-rural' (*Ibid*). Not all agriculture fully adopted the productivist regime, and elements of less productivist farming persisted in rural localities and in the everyday lives of the rural. The dynamics of rural depopulation, and later counterurbanization, impacted on rural localities presenting counter-narratives to the rationality of productivism. Equally, the formal representations of the rural as productivist were increasingly challenged by other formal representations of the rural contained, for example, in conservation policy and animal welfare discourse. Collectively, these tensions helped to corrode the productivist hegemony and led to the rethinking of British rurality from the 1980s onwards.

THE SCOPE AND STRUCTURE OF THIS BOOK

Keith Halfacree's three-fold model of rural space forms a useful reference point for introducing the ways in which this book proposes to engage with the idea of the 'rural'. The book aims to examine, explore and critique the different ways in which ideas of rurality and rural space have been produced, reproduced and employed in human geography and related disciplines. As such, it is interested not only in the representation of rurality and rural space within academic, policy, media and lay discourses, but also in the material effects of these ideas in rural localities, and in the ways in which these ideas are performed through the everyday practices of rural life.

In adopting this approach, the book is not intended to be a comprehensive survey of rural areas and their geographies. There are some aspects of rural life, and some areas of research within rural geography that are discussed only fleetingly in this volume (for more detailed discussion of these topics see Woods, 2005a, also Flora *et al.*, 2008, Ilbery, 1998). The narrative presented has also been shaped by other considerations. In order to emphasize current ideas and debates in rural geography, I have generally cited and referenced books and papers published in the last ten years, and consequently skipped over some 'classics' in rural geography that were important in shaping understanding of rural processes at the time, but which do not closely reflect present concerns. Similarly, as this book is part of a series on 'key ideas in geography', the narrative is primarily framed in terms of geographical research and literature, rather than work in rural sociology or agricultural economics, although in practice these disciplines are tightly interconnected and there is much that rural sociologists and economists will find to be of interest in the discussion.

This book also attempts to engage with the rural at a global scale, drawing on examples from both the 'global north' (the advanced industrialized countries of Europe, North America and East Asia, plus Australia and New Zealand) and the 'global south' (the economically poorer countries of Africa, Asia, Latin America and Oceania, also sometimes referred to as the 'developing world' or the 'third world'). However, it should also be acknowledged that this engagement is inevitably uneven because it is constrained by the spatial pattern of rural geography research. There are differences, for example, in the volume, focus and perspective of rural geography in Britain and in the United States (see Kurtz and Craig, 2009, and Woods, 2009b, for more discussion of this difference), not to mention between Anglophone rural geography and rural geography as practised in France, or Germany, or Japan, which will be influenced by different wider bodies of geographical literature. Moreover, research on rural geographies in the global north (conducted by 'rural geographers') has been largely divorced from research on rural geographies in the global south (conducted by 'development geographers'), with the latter tending to be more concerned with social and economic structures, and less influenced by the 'cultural turn' (see also Woods, 2009a).

The nine chapters of this book each focus on a different way in which geographers have engaged with the 'rural' as an idea and as an object of study. All the chapters have 'active' titles which are designed to emphasize the dynamic nature of our engagement with rural space and rural life. In some cases the active verb refers to the purpose which is being ascribed to the rural, as in 'Exploiting the rural' and 'Consuming the rural'. In other cases, it refers to the way in which the rural is brought into being, as in 'Performing the rural', whilst in others still it refers to the practices of researchers in constructing knowledge about the 'rural', as in 'Approaching the rural' and 'Re-making the rural'.

The next chapter, 'Imagining the rural', continues on from this chapter by examining in general terms the development of the 'rural' as an idea within both academic and lay discourses, and the materialization of these ideas through policy and practice. It extends the trajectory of conceptual development in rural geography, discussed in this chapter, to introduce the idea of a relational perspective on the rural. Together with Halfacree's three-fold model outlined above, this forms the foundation for the subsequent chapters, which each focus on a specific dimension of the rural.

The following three chapters hence concern constructions of the rural as an economic space. Chapter 3, 'Exploiting the rural', discusses the rural as a space of production for agriculture and other primary industries. In contrast, Chapter 4, 'Consuming the rural', explores the rural as a space of consumption, notably through tourism, which increasingly forms an alternative way of turning the rural into economic gain. Chapter 5, 'Developing the rural', considers the rural as an object of economic development, examining how development strategies feed on particular ways of conceptualizing the rural.

Chapter 6 and 7 shift attention to the lived experiences of rural space. Chapter 6, 'Living in the rural', discusses the meanings attached to rural lifestyles and rural communities, and the ways these are tested and remade through dynamic social change. Chapter 7, 'Performing the rural', moves away from representations of the rural, to consider the enactment of rurality through performance and the everyday practices of rural life, including 'more-than-representational' ways of knowing the rural. The final two chapters turn to more political perspectives. Chapter 8, 'Regulating the rural', discusses the governance of the rural economy and rural environment, and considers the representations of the rural that underpin rural policies and strategies of governmentality. Finally, Chapter 9, 'Re-making the rural', examines some of the key drivers of change in the twenty-first century countryside, the resulting political challenges, and the implications for the study of rural geography.

FURTHER READING

The three-fold model of rural space discussed in this chapter is developed by Keith Halfacree in his chapter on 'Rural space' in the *Handbook of Rural Studies* (2006). Also in the *Handbook of Rural Studies*, Paul Cloke's chapter on 'Conceptualizing rurality' critically reviews the functional, political-economy and social constructionist approaches to rurality employed in rural geography. For more on the history of rural geography as a sub-discipline, and on differences in the practice of rural geography between different countries, see the entry on 'Rural geography' by Michael Woods in the *International Encyclopaedia of Human Geography* (2009a). For more on the linguistic origins and development of the terms 'rural', 'country' and 'countryside' see Raymond Williams's classic book, *The Country and the City* (1973), and Michael Bunce's chapter in *Country Visions* (2003).

2

IMAGINING THE RURAL

INTRODUCTION

In the previous chapter, the 'rural' was revealed to be an elusive concept, a term that does not describe a hard, fast and indisputable material object, but rather refers to a loose set of ideas and associations that have developed over time and which are debated and contested. The rural sociologist Marc Mormont arguably put it best in referring to the 'rural' as a 'category of thought' (1990: 40), a description that emphasizes that the 'rural' is first imagined, then represented, then takes on material form as places, landscapes and ways of life are shaped to conform to the expectations that the idea of the 'rural' embodied. Experiences of these 'rural' places and lifestyles are fed back into the collective imagination, refining and modifying the idea and thus contributing to a dynamic process through which the 'rural' is produced and reproduced.

This chapter embarks on a more detailed examination of the production and reproduction of the rural as an idea, its imagination and representation in popular culture, and its translation into material form in the landscape. The chapter is structured in three sections. First, it traces the historical evolution of the idea of the 'rural'. It shows how ideas about the rural rapidly became detached from the physical spaces they were supposed to refer to, and follows the transportation of European ideas of the rural around with world and the modification of the concept as it was exposed to alien landscapes and environments. Second, the chapter questions how we 'know' the rural in the contemporary world, focusing on the representation of the rural through quantification, in the media, and through lay discourses. Third, the chapter considers the rural as a relational concept, exploring both the relations that constitute the rural and the shifting nature of the relation between the rural and the urban.

THE HISTORICAL CONSTRUCTION OF THE RURAL

Origins

The idea of the 'rural' has ancient origins. As soon as human populations began to concentrate in defensible settlements, so a need developed for a term to refer to the lands that lay outside these towns and cities. The earliest recorded root words for our modern term 'rural' appear to have done just that, referring literally to 'open space' (Ayto, 1990). By the time of classical Rome, more sophisticated representations of rurality had developed, not only conceiving of the countryside as a 'place', but also attaching to that place a series of moral and cultural associations. The Roman countryside was recognized as a source of food and natural resources, servants and soldiers, but it was also a political resource, where land could be redistributed as reward for military service. For wealthy Romans, a country villa was both a retreat and a status symbol; yet, rural migrants to the city were regarded as rough and uncouth.

These competing ideas of the rural were both recorded and romanticized in Roman literature, most notably in Virgil's *Eclogues*, written between 42 and 37 BCE (Short, 1991). Virgil, however, was influenced by the earlier writing of Theocritus, a Sicilian Greek, whose *Idylls*, written in the third century BCE, are widely regarded as the start of the pastoral tradition in Western literature (B. Short, 2006; J. Short, 1991; Williams, 1973). Both Virgil and Theocritus conveyed something of the harshness of rural life in their work – Virgil in particular recounting the plight of farmers and shepherds evicted from the land by political edicts – but in both cases their representation of the rural was overwhelmingly bucolic. Rural life was portrayed as simple, innocent and virtuous, revolving around the honourable profession of agriculture and involving closeness to nature (Williams, 1973).

Through this propagation of the pastoral myth, the sign of 'rural' began to become detached from the referent of rural geographical space over two thousand years ago. As the pastoral myth was reproduced within Western civilization over the centuries, so the ideas of the rural that it contained became entrenched in Western culture, and the dissonance of the sign and the referent was reinforced (Bunce, 1994; Short, 2006). Indeed, the perpetuation of the pastoral myth of the rural can be argued to have disguised crucial aspects of the condition of rural life throughout the mediaeval and early modern periods. It disguised the exploitation and oppression of the countryside, first through the brutality of feudalism and later through the appropriation of the rural land for capitalist production in the agrarian revolution. It also disguised the scale and complexity of connections between the city and the country, with the dynamics of rural localities and the everyday practices of rural life already heavily inscribed with more prosaic urban renderings of the rural: as a source of food, fuel and building materials, as hunting ground, and as a defensive buffer.

Furthermore, the pastoral representation of the rural competed in the urban imagination with starker representations of the rural as a place of danger and threat. This was the rural as wilderness, an idea that has even older origins than the pastoral myth, emerging with the development of agricultural societies around 10,000 years ago (Short, 1991). Fear of the wilderness runs through the folklore of mediaeval Europe, with tales of the evil spirits, monsters and savages resident in the forests, marshes and mountains. Wilderness could be tamed through cultivation, thus helping to boost the virtues of pastoralism, but even cultivated landscapes were perceived to harbour threats to urban travellers, in the form of vagabonds and highwaymen, extreme weather and the insularity of rural communities. Thus, there have always been at least two rurals in the urban imagination, each detached from the grounded everyday practices of rural life.

Globalizing the rural

European ideas of rurality were transported around the world from the fifteenth century onwards by explorers, settlers and imperial administrators, who both interpreted the new lands they encountered from the perspective of these ideas and attempted to recreate the European countryside in their new colonies. The early colonists arriving in North America, for example, were confronted with a land devoid of recognizable urban settlement, but which nonetheless required transformation before it conformed to the pastoral vision of European rurality. Although populated by native tribes, who farmed and harvested wild plants and animals, to European eyes North America was wilderness, and the conquest of the wilderness became a defining motif in the construction of American national identity (Short, 1991). The wilderness was tamed not through urbanization but through the replacement of an alien rural with a more familiar pastoral landscape (see also Box 2.1). As Knobloch (1996)

Box 2.1 GENDERING THE RURAL

Contemporary descriptions of the 'conquest' and 'taming' of the North American rural wilderness are highly charged with gendered metaphors that presented the land as female, in need of settling and enclosing through the masculinist assertion of the European pastoral tradition. Yet, as Kolodny (1975) pointed out, the patriarchal discourse of 'land-as-woman' contained an inherent conflict of meaning between maternal containment and sexual seduction. On the one hand, the land was imagined as a mother, 'whose generosity and abundance were marvellous, Edenic, but which could also overwhelm settlers and corrupt their efforts at self-sufficiency' (Rose, 1993: 105); but, on the other hand, the land was also represented as a temptress, inviting penetration. As such, 'domination of the land began to be seen as both incest and rape, and the horror of this necessitated a psychological and emotional separation from the land and from woman' (*Ibid*).

The feminizing of the American wilderness reflected a much older tradition of associating nature and women, and thus of representing the natural rural landscape as female. European and North American landscape painting has repeatedly portrayed sexually alluring women in natural settings, eliding the fertility of nature and the fertility of women. In some cases, female bodies were used to represent the landscape, such that 'the shapes of hills, the use of woods and flowers and the presence of water were included not only to symbolise the reproductive capacities of women but in the actual depiction of the female body' (Little, 2002: 52). This discourse positioned women as passive – embedded in the landscape – but men

continued

as active, the agents of transformation of rural space (Rose, 1993), in turn informing perceptions about the place of women in the countryside, including patriarchal ideas of landownership and the gendering of farmwork (discussed further in Chapter 7).

The dominance of the masculinist gaze over landscape has also been reproduced in the construction of the rural as a place of leisure and consumption (discussed in Chapter 4). As Rose (1993) has observed, the enjoyment of rural recreation has been equated with sexual pleasure, the aesthetic appreciation of rural landscape with the voyeuristic objectification of female beauty, and the exploration of the countryside by walkers, or cyclists or motorists with sexual discovery. The opposition of the passive female rural landscape to the active male consumer was further emphasized by the tendency of rural writing and guidebooks to portray tourists and recreationalists as men. Cover illustrations of mid-twentieth-century maps, for example, generally only allowed women to be glimpsed as car passengers, if at all, whilst men took centre stage as motorists, ramblers and cyclists.

Further reading: Little (2002), Rose (1993).

documents, agriculture was a tool of colonization in the American west, conducted according to European ideas and conventions. Husbandry practices, crops and livestock were all imported from Europe. Andalusian longhorn cattle, English Hereford cattle and Spanish merino sheep came to dominate grasslands cleansed of their native bison, even if some breeds struggled to adapt to climatic conditions and required 'improvement' (Knobloch, 1996). The cultivation of the American west appropriated the hunting grounds of native tribes and disregarded their knowledge of the land. Significantly, those tribes that did adopt European-style farming practices, including the Cherokee, Choctaw, Chichasaw, Seminole and Creek, were heralded as 'the five civilized tribes' (*Ibid*).

Yet, the agrarianization of North America did not produce a replica of the European countryside, but a new rural space that could be represented as an expression of American identity and virtue. As Valenčius (2002) observes:

> From the nation's founding, the connection between farming and virtue had found clear expression in American thought and writing. Orderly fields of straight rows clearly delineated from the surrounding unshaped terrain implied orderly virtue in their free cultivators. Apparently unfarmed land was a moral as well as a physical challenge. The confrontation with unfamiliar territory that took place in the canebreaks, the fields, and the prairies of the middle Mississippi Valley thus marks one of the defining aspects of American self-imagination, for good and for ill.
>
> (Valenčius, 2002: 10)

As the North American population became increasingly urbanized, these representations of rural America were idyllized and mythologized through art and literature. Bunce (1994) notes that at least 140 novels concerning farm life in the US Mid-West were published between 1891 and 1962, reproducing a rural idyll of agrarian simplicity, moral fortitude, innate wisdom and old-fashioned decency (see Box 2.2). Significantly, the rural

Box 2.2 THE RURAL IDYLL

One of the most powerful and enduring ideas about the rural is that of the 'rural idyll'. This imagines the rural to be a place of peace, tranquillity and simple virtue, contrasted with the bustle and brashness of the city. Whilst the rural idyll has also become associated with an escape from modernity, idyllic representations of country life are as old as writing about the rural, and in each historical era people have embellished the rural idyll with antonyms to their own apprehensions (Short, 2006). Representations of the rural idyll were particularly popularized during the late nineteenth and early twentieth century, as Europe and North America became increasingly urbanized and industrialized. The rural idyll fed on discourses of anti-urbanism, agrarianism and nature that were used to differentiate between the urban present and a romanticized rural past, particularly by nostalgic urban residents. Bunce (1994) describes this as the 'armchair countryside', imagined and appreciated from urban and suburban sitting rooms. As such, the 'rural idyll' has commonly been an idea imposed on rural areas and communities from the outside.

There are in practice many different rural idylls, with different cultural and moral emphases and different pictorial representations. For example, ideas of the visual manifestation of the rural idyll vary by region and nation – from the rolling downland of England to the forests and lakes of Scandinavia – and are often closely tied to ideas of national identity. In Anglophone countries, ideas of the rural idyll are strongly influenced by a romanticized and nostalgic memory of pre-industrial rural England, even if the landscape features this idea conveys are alien to large parts of North America or Australia. Nonetheless, Hollywood has played a key role in reproducing ideas of the rural idyll in the modern era, contributing to its global diffusion – perhaps ironically, given that the rural idyll itself presents rural communities as detached and sheltered from the pressures of global interconnections (D. Bell, 2006).

In addition to film, the rural idyll is reproduced through art, literature, poetry, music and television, and is an idea that is pervasive in popular culture. Yet, the rural idyll can have real material effects, as later chapters in this book discuss in more detail. The attraction of the rural idyll is a major pull-factor in counterurbanization and a selling point for rural tourism (see Chapter 4), but because rural landscapes and lifestyles so often fail to live up to the image of the rural idyll, they need to be modified to meet the expectations of investors and customers. The concept of the rural idyll has also influenced political ideas and government policies – for example in the enduring symbolic significance of the 'family farm' – whilst disguising some of the harsher realities of rural life, including poverty, prejudice and environmental problems (see Chapters 6 and 8). The rural idyll is therefore a normative concept, in that it seeks to construct rurality in a certain way rather than representing the rural that actually exists. It is in this sense that Halfacree (1993) has argued that the rural idyll is about the visioning of rural areas by a hegemonic middle-class culture, and Bunce (2003) has noted that 'even if we accept that there are many versions of the rural idyll, they all converge around a normative nostalgic ideal which is embedded in social and economic structures' (p. 25).

Further reading: Bunce (2003), Short (2006).

of these representations is defined not only by a landscape, but also by the everyday practices of rural residents. These practices were both studied and promoted by the nascent academic discipline of rural sociology in the United States (M. Bell, 2007; Woods, 2005a), yet as with the European pastoral myth, the idea of the virtuous American yeoman was even by the early twentieth century divorced from the materiality of an increasingly industrialized agriculture (see Chapter 3).

Alongside eulogizing the agrarian idyll, nineteenth-century American culture embarked on a re-appraisal of the wilderness. Writers such as James Fenimore Cooper, John W. Audubon and George Catlin, and painters such as Albert Bierstadt and Thomas Moran, popularized a new representation of the American wilderness as a place of natural beauty and scientific and spiritual significance (Short, 1991). Audubon's comparison of the wilderness to the Garden of Eden, in particular, encapsulated both the romanticization of the wilderness and its elevation to a symbol of permanence and endurance. The establishment of the first 'National Parks' in areas of wilderness such as Yellowstone reflected both their representation as places of sublime landscapes and their identification as expressions of a distinctive American history that could rival the crowded cultural history of Europe (Runte, 1997). Yet, as Cronon (1996) observes, the re-invention of the American wilderness by mainly urban protagonists largely ignored the historical presence of Native Americans, noting that 'the myth of the wilderness as "virgin", uninhabited land had always been especially cruel when seen from the perspective of the Indians who had once called that land home' (p. 15).

The making of the American countryside was hence a process of hybridization, combining European ideas and materials with the native landscape, flora and fauna, and fusing cultural and political ideals. Similar concoctions were constructed in other regions subjected to European colonization. In Canada, rural settlement followed a comparative moral and cultural imperative, providing a justification for the displacement of native peoples and reordering of the landscape along European lines: 'the agricultural landscape had a powerful symbolic meaning, inscribing Englishness into the very hills, emphasizing European possession of the land, and enabling settlers to imagine themselves as recreating a version of England in a new place' (Murton, 2007: 11). Murton particularly records the rationalities of rurality that underpinned large-scale land resettlement programmes in southern British Columbia in the early twentieth century.

The schemes, which involved draining lakes and marshes, constructing canals and irrigating plains and were aimed at settling First World War veterans into wholesome rural livelihoods, were promoted through imagery that owed more to imagination than the reality of the locality, as Murton describes for the cover of a brochure produced by the Land Settlement Board:

> The illustrator envisions a patchwork of fields receding into the distance, shrubbery between the fields suggestive of hedges. Tidy houses dot the scene. The entire landscape is framed by the peaks of the Coast Mountains, which pierce the clouds that shroud the Fraser Valley. The result: a lovely (if somewhat incongruous) amalgam of BC and England – the rugged west and the pastoral old country – an imagined geography uncomplicated by reality.
>
> (Murton, 2007: 2)

Yet, as Murton proceeds to detail, the countryside under construction was a *modern* countryside. It was achieved through the application of new technologies and demonstrated the improvement of nature. Moreover, the land resettlement programmes were framed politically by the ideology of Canadian liberalism. The landscape of small farms reflected a belief in individual property and enterprise as the foundation of economic prosperity and social order; whilst the scale of the projects was facilitated by the growing involvement in inter-war Canada of the state in social and environmental reform. Ultimately, however, Murton argues that the modernizers were constrained by their ties to an English idea of the rural and missed the opportunity to really harness the natural resources to them. They were focused on 'converting existing landscapes into what they thought of as the proper countryside' (*Ibid*: 195), but their agricultural ideas fitted poorly with the environmental conditions of British Columbia.

In New Zealand, too, the colonial countryside had a contested and experimental evolution. During the late nineteenth century, the open country of New Zealand was transformed from wild grassland to pasture, with the area of improved pasture increasing from 158,000 acres in 1861 to 16.5 million acres in 1925, supporting an agricultural economy that was dependent on exporting wool to Britain, using sheep breeds introduced from Britain and raised using British husbandry (Holland *et al.*, 2002).

Indeed, almost all of the key components in the construction of the colonial countryside were imported:

> From the outset, settlers brought in plants and animals native to temperate and sub-tropical regions of the world. Trees and shrubs, herbs and grasses, flowers and vegetables, birds and mammals, game and economic animals, insects, and fish were imported in bulk and widely distributed. After a few years, when native grasses had largely failed to meet pastoralists' expectations, introduced grasses and herbs were either purposefully sown or accidentally introduced as contaminants in the widely used 'station seed mix'.
>
> (Holland *et al.*, 2002: 75)

Some imports failed to behave in the way anticipated, notably rabbits which by the 1880s had been declared the number one nuisance for agriculture, as well as causing extensive damage to native flora. The impact of the constructed countryside was not only environmental but also cultural, displacing Māori for whom the open country had been 'a mosaic of productive aquatic and dryland habitats for fibre, food plants, and animals' (*Ibid*: 71). Instead, the new rural New Zealand became the basis of a distinctive settler Pakeha identity (C. Bell, 1997). As in North America, the identification of the rural as the 'real' New Zealand reflected not only a discourse of the virtues and simplicity of rural life, but also a belief that the struggle to 'break the land' had helped to forge the national character (*Ibid*). Only recently has this representation been challenged by the re-assertion of Māori narratives of the rural landscape and the plants, animals and resources that it contains (see Panelli *et al.*, 2008).

Colonial settlers in tropical and sub-tropical regions were confronted with a landscape even more removed from that of the European countryside, but here too hybridization was an intrinsic feature of empire, combining European ideas and expectations of the rural with the local environment (Casid, 2005). Colonial administrators raised on English country estates attempted to recreate English country gardens in India and the West Indies (Roberts, 1998), and transported rituals of English rural life such as cricket and fox hunting. By the early twentieth century, some aristocrats alarmed at the pace of social and political reform in Britain were emigrating to reconstruct their country estates in South and East Africa.

Even the commercial crop plantations that owed little to the tradition of European agriculture were infused with motifs of European rurality, as Jill Casid describes in the West Indies:

> The colonial hybrid landscape produced was a hybridization of the island of Jamaica not only with Italy and Tahiti but also with the rural farmscapes of the Netherlands and of England. A detail from a 1756 map of Jamaica by the island engineer Thomas Craskell and the surveyor James Simpson represents a picture of the mule mill, windmill, cane storehouses, and divided cane fields of a sugar plantation. With the exception of the topographical feature of the mountainous terrain in the background and the black slave driving the cattle-drawn wagon, the over-all composition, winding lane, and vegetation are rendered to make the Jamaican sugar plantation as much like a fantasy of rural England and the Netherlands as possible.
>
> (Casid, 2005: 70)

Critically, though, the transportation of 'rural' objects and diffusion of rural ideas was not all one-way, but also flowed back from the colonies to Europe. Crops such as potatoes and maize were introduced to Europe from America, becoming staples in some regions and transforming the economic structures and everyday practices of rural localities. As Casid records, wealth generated by colonial plantations and imperial trade was also frequently transported back to Europe and paid for the construction of extravagant country estates. The physical transportation and inclusion in these parklands of trees and shrubs such as rhododendrons, magnolias and azaleas served as a reminder of the source of the landowner's wealth and became incorporated into the European rural picturesque.

Countryside preservation and the time-spaces of the rural

The rural landscapes and societies of Europe that inspired and informed colonial settlers were by the end of the nineteenth century coming under increasing pressure from urban expansion. This was particularly the case in Britain, where half of the population was already living in urban areas by 1851 and that proportion increased steeply to four-fifths by 1951 (Saville, 1957). The depopulation of the countryside reflected the disjuncture between the idyllic representation of rural life and the realities of everyday experience. People moved to towns to escape the poverty, isolation and lack of social mobility of the countryside, hoping for better living conditions and opportunities for advancement (Short, 2000).

Yet, the dominant cultural response to urbanization was a renewed popularization of idyllic representations of rural Britain. Writers such as William Morris (*News From Nowhere*, 1891) and, later, Clough Williams Ellis (*England and the Octopus*, 1928) revisited traditional associations of the rural with virtue, innocence and simplicity, but employed these as ammunition in a new anti-urban and anti-industrial sentiment. In contrast to some earlier representations, the rural was portrayed as fragile, vulnerable to urban incursions, either physical (the disorderly encroachment of suburbs and ribbon development), or socio-cultural (the loss of rural traditions and crafts to the diffusion of urban culture). These sentiments appealed to a new middle class that was already developing interests in nature study and outdoor recreation, providing the foundations for a preservationist movement that coalesced around organizations such as the National Trust (founded 1895) and the Council for the Preservation of Rural England (CPRE) (founded 1926) (Bunce, 1994; Short, 2006).

The idea of rurality propagated by the British preservationist movement involved two key features. First, it was closely tied to national identity. The association of rurality and national identity in Britain can be traced to at least the fourteenth century, but the coincidence of the perceived threat to the countryside and the threat of war at the start of the twentieth century created a sense of shared destiny. Bucolic rural images were used as recruiting and morale-boosting posters in the First World War, but preservationists claimed that the rate of urbanization during the war years betrayed this message. A famous cartoon in the magazine *Punch* depicted 'Mr Smith' leaving a small village in 1914 to answer 'the call to preserve the native soil inviolate', and returning in 1919 to a burgeoning city (Matless, 1998). The association was reinforced in the 1920s and 1930s in popular guidebooks and travelogues with titles such as *In Search of England* and *The Legacy of England* (Brace, 2003). At the launch of the CPRE in 1926, founding secretary Professor Patrick Abercrombie made clear that preserving the countryside was the same as preserving England:

> The greatest historical monument that we possess, the most essential thing which is England, is the Countryside, the Market Town, the

> Village, the Hedgerow Trees, the Lanes, the Copses, the Streams and the Farmsteads.
>
> (Patrick Abercrombie, quoted by Lowe and Goyder, 1983: 18)

Second, the preservationist movement demanded the orderly separation of the rural and the urban. In spite of the rhetoric of urban–rural differences, the actual boundaries (physical and imagined) between town and country had always been fluid. Attempts to impose a stricter spatial ordering were first driven from within the city, with the expulsion of livestock and abattoirs on sanitary grounds (Philo, 1995). The preservationists sought to constrain the city from without. Matless (1998) describes this desire for spatial order as a 'morality of settlement' (p. 32), with urban and rural kept in their appropriate place. He again quotes Abercrombie as an illustration of this discourse:

> The essence of the aesthetic of Town and Country Planning consists in the frank recognition of these two opposites ... Let Urbanism prevail and preponderate in the Town and let the Country remain rural. Keep the distinction clear.
>
> (Abercrombie, quoted by Matless, 1998: 32)

The fruit of these efforts was the 1947 Town and Country Planning Act, which introduced development controls, drew envelopes around settlements and created 'greenbelts' around major cities in which new development was prohibited. The effect was to achieve the 'containment of urban England' (Hall *et al.*, 1973). Yet, as Murdoch and Lowe (2003) observed, almost as soon as the separation of urban and rural was imposed, it was transgressed by socio-economic processes. In what they call the 'preservationist paradox', the 'implementation of preservationist policy ensures a continuing transgression of the divide' (*Ibid*: 328). Villages protected from unrestrained development became attractive to urban migrants seeking the 'rural idyll', who in turn became supporters of groups such as the CPRE in order to halt further transgression.

The problems of imposing a spatial order on town and country at an arbitrary point in time, together with the strong strain of nostalgia in preservationist representations of the rural, might lead us to consider the 'time-spaces' of the rural. In other words, as well as asking '*where* is the rural?', we might also ask '*when* was the rural?' Representations of the rural commonly construct the rural in terms of a link to the past. Rural people are venerated as national icons because they are believed to retain traditions and practices handed down from tribal ancestors; whilst idyllic representations position the rural as a refuge from the pressures of modern life (J. Short, 1991) (Plate 2.1).

Yet, fixing the time of the rural is impossible. Matless (1994) observes, for instance, that it 'seems often to be assumed that the English village lies on the side of tradition against modernity, with those two terms in opposition' (p. 79), but notes that the villages represented by preservationist writers such as W.G. Hoskins and Thomas Sharp do not fit this simple dualism. Equally, whilst the protection of historical buildings is a major aspect of countryside preservation, these do not date from a consistent period. Rural literature alludes to a 'golden age', but this elusively slips further and further back (B. Short, 2006).

Nonetheless, the commodification of heritage is a central part of the practice of present-day rural localities, not only in Europe but also around the world (see C. Bell, 1997; Crang, 1999; Prideaux, 2002; Wilson, 1992).

Plate 2.1 Thatched cottages and cornfields: the English rural idyll? (Photo: author)

In order to correspond with modern expectations of the historic rural, the presentation of rural heritage can be subject to revisionism and embellishment, processes that can be controversial. Rigg and Ritchie (2002), for example, describe the commodification of rural heritage in Thailand, a country in which the interpretation of the rural past is politically sensitive. Idyllic representations of the rural past have been employed by radical activists to challenge present-day modernization programmes and assert a peasant-based history. The same idyllic representations have also, however, been appropriated by new tourist resorts, the development of which displaces peasant populations. The result is an imagined performance of 'authentic' Thai rural life that has little to do with the reality of rural Thailand, either past or present:

> [The resort] shows how urban Thais conceptualize the rural, as the owners and designers (both Thai and *farang*) designed the 'rural' experience for the consumption of other – a mythic past performed as a pageant (much like the laser light shows of 'ancient' Sukhothai) – with little connection to the reality of subsistence rice farming.
>
> (Rigg and Ritchie, 2002: 367)

KNOWING THE RURAL

Discourses of rurality

The contemporary rural is complex space, created by the diverse and dynamic processes of imagination, representation, materialization and contestation described in the section above, and taking on different forms in different contexts and from different perspectives. The process of reimagining and reproducing the rural is necessarily reflexive, it involves critical reflection on the observed state of the 'rural' and comparison with some imagined yardstick of what we might expect the rural to be like. As such, the reproduction of the rural requires us to have some way of making sense of the rural, of *knowing* the rural. Thus, whilst the previous section focused on the ontology of the rural – how the rural has been brought into existence – this section will examine the epistemological question of how we know and understand the rural.

Indeed, there is not one singular way of knowing the rural, but rather there are in circulation a multitude of different *discourses of rurality*, which each represent a different way of knowing and understanding the rural. The term 'discourse' is used here in a sense that is derived from the work of Michel Foucault, where a discourse is a way of understanding the world, or rendering visible certain relationships, practices and subjectivities that constitute a framework of knowledge, a 'vast network of signs, symbols, and practices through which we make our world(s) meaningful to ourselves and others' (Gregory, 1994: 11). A discourse therefore is not just a representation of reality, it creates reality by producing meaning and setting the boundaries of intelligibility. Neither does a discourse consist solely of written and spoken words, it also includes images, sounds, bodily actions such as gestures, habitual thoughts and practices, and so on (Wylie, 2007). Thus, to take an example from the previous section, the association of American rural life with simplicity, moral fortitude, wisdom and decency was a discourse constituted not only by written texts, but also by art, music, political speeches and church sermons, and by the everyday practices of rural dwellers that conformed to these ideas.

Academic knowledge about rural space that is produced by geographers is a discourse (or, rather, multiple discourses as academic knowledge is far from coherent and consistent) (Halfacree, 1993), but the production of discourse is not the preserve of academics. We can also identify *political* and *policy discourses of rurality* that enable the state to know the rural and frame the governance of rural space; *media discourses of rurality* that frame both 'factual' and 'fictional' stories about rural life and serve to popularize ideas of the rural; and, not least, *lay discourses of the rural*, that are constituted by the beliefs, thoughts, descriptions and actions of ordinary people in their everyday lives.

This section concentrates on three particular aspects of 'knowing' the rural. First, the construction of knowledge about the rural through quantitative data as part of academic and governmental discourses; second, the role of media discourses in framing popular understanding of rural life; and third, the engagement of lay discourses of rurality in constructing geographical knowledge about the countryside.

Quantifying the rural

Historically, both academic and governmental discourses of the rural adhered to a positivist epistemology which held that the rural could be accurately captured in quantitative data and processed to produce a 'true' representation of various aspects of rural space and rural life that could be statistically tested and validated. By the same token, non-numerical representations of the rural were dismissed as unreliable and irrelevant as they could not be tested and validated in the same way. Quantitative methods hence dominated the more analytical branches of rural geography from the 1960s onwards (there also persisted a strong tradition in rural geography based on field description, especially the North American school associated with John Fraser Hart at the University of Minnesota). Quantitative analysis could be employed to produce knowledge about almost any aspect of the rural. Thus, knowledge and understanding of agriculture was constructed through the analysis of a vast array of data relating to crop production, field sizes, fertilizer application, farm incomes, farm size and so on. Similarly, knowledge and understanding of rural settlement patterns and settlement function was constructed through quantitative analysis of population data, measurements of distance between settlements, and enumerations of service provision, combined with a range of socio-economic data.

It was a logical extension of this approach to attempt to use quantitative methods to define and delimit rural areas, either based on specific indicators such as population, population density, accessibility, land use or agricultural employment, or through the development of more sophisticated multi-indicator models (see for instance Cloke, 1977; Cloke and Edwards, 1986). However, far from producing a 'true' map of rural areas, such models instead generated a large number of different, over-lapping yet non-congruent, rurals depending on the indicators selected and the scale of the territorial units used. More broadly, the movement in rural geography towards first political-economy and later cultural approaches started to raise questions about the assumptions underlying quantitative analysis and usefulness of the knowledge produced. In so doing, such critiques revealed the quantitative approach to be just one particular discourse, but a discourse that had set the boundaries of knowledge production in rural geography, as Paul Cloke acknowledges in a later reflection on his early 'indices of rurality' for England and Wales:

> Given my view now that this work is an inappropriate way of addressing the idea of what and where is rural, I have often asked the question of why I did this indexing. My empirical work on evaluating key settlement policies was focusing on parts of Devon (which I constructed as a 'remoter' rural area) and Warwickshire (a 'pressured' rural area)

> and so although I persuaded myself otherwise, the index was not necessary for selecting case study areas. Apart from the 'prevailing social science culture' which legitimized and maybe even necessitated this sort of thing, I can only suggest that I was expressing a rather naïve interest in the question of what 'rural' was/is in the only way that at the time I had the academic and cultural competence so to do. I think that I knew at the time that by selecting a number of variables to represent, collectively, the rural I was pre-determining the outcome, but the interest was in the emerging geographies of that pre-determination.
>
> (Cloke, 1994: 156)

Two caveats must, however, be attached to the potted account of the demise of quantitative approaches given above. First, the move away from quantitative methods in rural geography has been much stronger in Britain than, for example, in the United States or in much of Europe, where the majority of rural geography research continues to involve at least some quantitative analysis. Second, an emphasis on quantitative ways of knowing the rural is still prevalent in governmental discourses. This is evident in the regular collation and publication of official statistics relating to diverse areas of rural life, from agricultural production to healthcare provision, which are in turn used to inform policy decisions. The Commission for Rural Communities in England, for example, produces an annual 'State of the Countryside' report that presents a statistical digest of the social, economic and environmental condition of rural England, which by aggregating data to the level of 'rural England' reproduces the discourse that the English countryside exists as a singular, coherent entity.

Moreover, the need for governments to demarcate the territories to which rural policies apply has perpetuated the use of quantitative-based definitions of rural areas. Once again, the tailoring of definitions to specific policy sectors has produced a proliferation of different non-congruent definitions. Shambaugh-Miller (2007) notes that there are over 50 different definitions of rural areas used by federal programmes in the United States, with considerable discrepancies in their spatial coverage (see also M. Bell, 2007). For example, there are over 30 million people who are defined as 'rural' according to the census definition, yet who live in areas categorized as 'metropolitan' by the definition used by the Office of Management and Budget. The detail of these definitions can have material effects in rural localities as they effect eligibility for funding and the implementation of other programmes. Shambaugh-Miller, for instance, highlights the case of access standards for Medicare pharmacies, which were intended to ensure that 70 per cent of the rural population of the US lived within 15 miles of a Medicare pharmacy. However, the definition of rural area used in the programme was so broad that it included everywhere apart from core inner cities, such that the 70 per cent threshold was easily met in suburban areas without increasing accessibility in remoter rural areas.

In 2002, the British government commissioned the development of a new 'rural definition' for England and Wales which was intended to provide a common classification for use by government departments and agencies. The initiative was at least in part politically motivated, reflecting the government's desire to construct an 'objective' representation of rural areas in order to counter the subjective representations mobilized by pressure groups such as the Countryside Alliance, which portrayed a countryside defined through traditional activities such as hunting and farming (Woods, 2008a). The new rural–urban classification was constructed by applying GIS technologies to the analysis of data on household density at a variety of scales (Bibby and Shepherd, 2004). However, the model produced by the exercise illustrates the continuing contingency of any quantitative-based definition of rural areas in two ways. First, the model actually stops short of labelling localities as being rural. Rather it positions territorial units within a two-dimensional matrix, with one vector representing settlement form, from urban to 'hamlets and dispersed households', and the other vector representing the regional context as either sparse or 'less sparse'. Communities with a population of more than 10,000 are considered to be 'urban' regardless their household density. Second, as the model is constructed at a number of spatial scales, a single point can be classified differently depending on the scale at which it is viewed. Thus, whilst the new classification has produced a more nuanced representation of rural England and Wales, it also demonstrates that there is no such thing as an objective definition of rurality.

Media discourses of rurality

The media, in its broadest sense, has always played an important part in the propagation and dissemination of ideas of rurality. The poems of

Theocritus and Virgil in the classical era were early media representations of rurality, as were later references to the rural in art, literature and song. The significance of the media in framing and reproducing discourses of rurality has been intensified since the nineteenth century by the coincidental rise of the mass media alongside the rapid urbanization, such that the media is now the means by which the majority of the population comes to 'know' the rural. This includes both apparently 'factual' accounts of rural life in the news media and documentaries, and 'fictionalized' stories of the rural narrated through literature, film and television dramas – although, as will be discussed below, there is slippage between the two modes.

Whilst the media can create and promote new discourses of rurality, media representations of the rural more often reflect and consolidate existing discourses of the rural, often inflected with particular moral, cultural or political beliefs. Indeed, the media can help to reproduce discourses of rurality as much through silence as through active representation. The mainstream news media, for example, is frequently accused of neglecting rural issues, thus reproducing the idea of the countryside as a quiet, tranquil, even backward place where nothing interesting happens (Harper, 2005). Rural news appears only to feature when it concerns events that are discordant with dominant discourses, and gets reported in language that evokes imagery of the rural idyll, as Bunce (2003) observes with respect to news reports of a fatal *E. coli* breakout in a small Canadian rural community.

Significant coverage of rural issues occurs only at times of apparent 'rural crisis', such as during the foot and mouth epidemic in Britain in 2001, or the more recent concerns about global food security. These events attract attention because they demand a reappraisal of discourses of rurality, and in such circumstances the news media can provide an important function in shaping and hosting debates on rural futures. Juska (2007) explores one such instance in an analysis of rural reporting in the *Lietuvos Rytas* newspaper in post-Soviet Lithuania. Juska records that the frequency of rural stores increased more than ten-fold between 1991 and 1999, as rural areas adjusted to economic liberalization and the implementation of land reforms. However, he also observes a shift in the discourse through which rural reports were framed. During the 1990s, Juska argues, reporting of rural Lithuania was largely unsympathetic, reflecting urban prejudices about the 'backwardness' of the rural population, support for radical reforms, and resentment at the dependency of the 'new' rural poor left behind by reforms on state welfare.

By 2000, however, the Lithuanian economy had strengthened and the country was preparing for entry into the European Union. In this context Juska identifies a shift in the dominant media discourse of rurality. Not only did the number of reports concerning the rural fall sharply, but 'Lithuanian rurality, instead of being defined as a welfare problem financed by urban classes, was increasingly reframed as an EU-wide problem' (2007: 251). As such, rural areas were now represented as an arena for rural development, with opportunities for new non-agricultural enterprises and lifestyles. In this way, the discourses of rurality reproduced through the Lithuanian news media both reflected popular perceptions of rural areas, and contributed to establishing the context for rural political reform.

A still greater influence in shaping popular ideas of the rural is exerted by the entertainment media, especially films and television drama series with rural themes, the potency of which is enhanced by their combining of narrative, sound and image, and by their reach and popularity. Phillips *et al.* (2001), for example, noted that rural dramas are amongst the most widely watched television genres in Britain, including programmes such as *Heartbeat*, *Peak Practice*, *Emmerdale*, *Midsomer Murders* and *Ballykissangel*. These programmes present a stylized and exaggerated version of the rural that is detached from the everyday material experience of rural life, yet which form a crucial means by which millions of viewers come to know the rural. In many cases, the appeal of rural dramas might be explained by their reinforcing of urban perceptions of the rural idyll. This holds even for the significant number of crime dramas that fall into this category, where the incongruity between the crime and the idyllic setting forms part of the mechanics of the programme. Aspects of the rural idyll are reproduced through plots, characterizations and scenery, as well as through a number of narrative devices: several popular rural programmes are period dramas, playing-up the nostalgic element of the rural idyll; rural programmes frequently introduce an 'outsider' character to highlight urban–rural differences; and, as Phillips *et al.* (2001) discuss, many programmes implicitly portray a countryside that is structured by class yet free of class conflict, with altruistic professionals and world-wise but content working classes.

There is, however, another rural to be found in films and television programmes, which feeds on urban prejudices about the insularity and 'backwardness' of rural communities and on primeval fears of the wilderness. This 'anti-idyll' forms a strand through films such as *Deliverance*, *The Wicker Man* and *The Blair Witch Project* (D. Bell, 1997), as well as (in parodied form) the British television comedy series, *The League of Gentlemen*. The fact that both strands of rural programming can sit alongside each other in television schedules testifies to the discriminating and critical nature of television audiences which are able to understand the rural presented in programmes as an exaggerated and fictionalized entity (Phillips *et al.*, 2001). Yet, viewers still make associations between the rural portrayed in television programmes and rural places that they have visited, and pick up elements of 'rural knowledge' from watching such programmes.

The discourses of rurality reproduced through films and television programmes can therefore have material effects in rural localities in three ways. First, the global dissemination of 'rural' films and television programmes has contributed to the detachment of representations of rurality from actual rural places. Urban viewers whose impressions of the rural are received through the media rather than through direct experience of rural areas assemble a composite idea of the rural informed by programmes set in a range of different rural environments, and may in turn apply these ideas to their local countryside (Phillips, 2008). Phillips (2004), for example, observes that the popularity of British rural television dramas in New Zealand means that 'images of the British countryside may hence contribute as much to New Zealanders' sense of the rural as do images of the New Zealand countryside' (p. 27).

Second, some rural campaigners and commentators have argued that media representations of the rural idyll combined with anthropomorphic representations of animals in film, television and literature have promoted a 'false' image of rural life and shifted public opinion on issues such as hunting and farming practices. A magazine feature for the Countryside Alliance pressure group in Britain, for example, asserted that 'a generation brought up on *The Animals of Farthing Wood*, Walt Disney films and visits to theme parks is easy meat for single-issue pressure groups who exploit this lack of understanding of the realities of the countryside to their own ends' (Hanbury-Tenison, 1997: 92).

Third, the filming locations of many films and television series have become tourist sites attracting programme devotees searching for some essence of the fictional rural represented. In some cases, the actual landscape has been modified to incorporate elements of the fictional setting.

Mordue (1999), for example, describes the case of the Yorkshire village of Goathland which experienced a surge in tourist visits after it became the filming location for the popular drama *Heartbeat*. As a result, the village landscape has become a hybrid of the actual community of Goathland and the fictional community of Aidensfield, with elements of the film set retained, such as the frontage of the village shop in the programme.

Lay discourses of rurality

Lay discourses of rurality are defined by Owain Jones (1995) as 'all the means of intentional and incidental communication which people use and encounter in the processes of their everyday lives, though which meanings of the rural, intentional and incidental, are expressed and constructed' (p. 38). Lay discourses are informed by, and provide feedback into, media discourses, policy discourses and academic discourses, but they are different in that they are grounded in the everyday practice of life in the countryside. Thus, as Jones observes, 'lay discourses of the rural amount to a significant part of rurality itself, directly shaping and mutating it, and also interlinking with other levels of discourse which have their own impacts' (1995: 39). Lay discourses accordingly attempt to describe and explain the 'everyday lives of the rural' in Halfacree's (2006) three-fold model of rural space, introduced in Chapter 1, interacting with the media, policy and academic discourses that comprise the 'formal representations of the rural'.

This is not to say, however, that lay discourses of rurality are any more 'authentic' than other discourses. They are 'situated knowledge', produced by the experiences of individuals and reflective of their age, gender, ethnicity, social class, education, residential history and so on. The value of studying lay discourses to rural geography is that they tell first-hand stories of rural life. They can assist us more than any other form of representation in understanding what people in rural areas think about rurality, and examining the connections between knowing the rural and acting in rural space. This can be illustrated by briefly focusing on four key elements of lay discourses of rurality.

First, lay discourses of rurality articulate how people in rural areas understand their locality to be rural. Significantly, these perceptions draw on popular discourses circulated through the media but connect them to actual experiences. Thus, two of Jones's respondents in southern England implicitly reproduce a discourse of the rural as agricultural in recounting experiences of being held up by tractors and livestock on the local roads:

> We are fortunate to have several local farms, animals graze the fields. Tractors track up and down the road. Not always a blessing!
>
> We regularly get stuck behind the cows on their way to or way back from milking. We hear sheep, birds, tractors, etc.
>
> (quoted in Jones, 1995: 42)

Second, lay discourses of rurality can set boundaries of who is, and who is not, considered to be rural, and thus act as exclusionary devices. Bell (1994), for example, in his ethnographic study of an English village describes how locally born residents distinguished themselves as 'country people' from in-migrants by way of certain knowledge about the countryside and certain practices and traditions. One interviewee, for instance, told Bell that her grandmother always said that she could not be a real country girl until she learned to eat 'jugged rabbit', or rabbit cooked in its own blood. Similarly, the lay discourses of rurality held by urban residents can reinforce stereotypes about rural people.

Third, lay discourses convey understanding of change in rural communities, including perceptions that places are becoming less rural. This is evident both in the work of Bell (1994) and Jones (1995) in England and in a report for the W.K. Kellogg Foundation on perceptions of rural America:

> Very few of its (the village) people work in agriculture so it is not as real as it was 20 or 30 years ago.
>
> (quoted by Jones, 1995: 42)

> Mostly it's cities moving out. And then you're not rural American anymore ... that's a tough one. The people coming in, people selling their land and people coming in.
>
> (quoted by W.K. Kellogg Foundation, 2002: 14)

Fourth, lay discourses frame the lived experience of rural life and the perceived needs of rural residents. In some cases, lay discourses reproduce ideas of the countryside as a safe, peaceful and self-supporting place to live associated with the rural idyll; yet, lay discourses can also challenge these comfortable assumptions with stories of poverty and prejudice. These accounts and tensions are explored further in Chapter 6.

RELATIONAL RURALS

Reflecting on rural relationality

The preceding sections of this chapter have highlighted a number of features of the idea of the rural: that it is a social construct, brought into being as it is imagined; that as a social construct, the rural is commonly imagined in relation to another entity (the rural is *closer* to nature, *less* developed, *more connected* to the past, etc. than the urban); that the social construction of the rural does not occur in a vacuum, but is embedded in networks of social, economic and political relations; that representations of the rural have become detached from the referent of rural space, and have been disseminated and transported through global networks; that the idea of the rural is materialized in rural localities through social, economic and political relations involving a variety of actors, both human and non-human, indigenous and exogenous.

As is apparent from this brief summary, a recurrent motif in our approach is the significance of relations. Indeed, we might be said to be constructing a relational account of the rural. The 'relational approach' has become popular in Human Geography over the last decade or so, and may be simply described as an emphasis on the significance of networks, connections, flows and mobilities in constituting space and place and the social, economic, cultural and political forms and processes associated with them. The relational approach rejects concepts of space and place as fixed entities, constrained within the static and hierarchical architecture of territory and scale (Marston *et al.*, 2005), and instead positions space as 'a product of practices, trajectories, interrelations' (Massey, 2004: 5), forever dynamic and contingent. As Wylie (2007) observes, 'relations do not occur in space, they *make* spaces – relational spaces, and the geography of the world is comprised of these' (p. 200).

Moreover, the relational approach retreats from the privileging of the social that had been a characteristic of Human Geography. It collapses the dualisms of nature and society and human and non-human, adopting an agnostic position of regarding all entities, human and non-human, as equal components within a network, each with a capacity to change outcomes through their participation or non-participation (see also Murdoch, 1997a, 1998, 2003; Woods, 1998a). Adopting this perspective involves trying to capture actions and impulses which by their nature cannot be represented. As such, the relational approach is also concerned with the 'non-representational' – emotions, impulsive bodily actions, affective relations (Thrift, 2007).

Murdoch (2003) argued that the relational approach presents an opportunity to resolve a tension that has existed in rural geography perspectives on the rural between the social construction approach and an equally prevalent recognition of the countryside as a place of engagement with nature. The second perspective, he suggested, limited the first, that 'the idea that the countryside is simply a social construction, one that reflects dominant patterns of social relations, cannot adequately account for the "natural" entities found within its boundaries' (p. 264). Murdoch proposed that the erasure of the nature–society divide in relational theory (and especially in Actor Network Theory), provided the foundation for a new approach, that 'combines the "social" and the "natural" perspectives with which people are familiar in order to treat the countryside as a *hybrid* space, one that mixes up social and natural entities in creative combinations' (*Ibid*).

Hybrid rurals

The concept of hybridity, as it is employed here, is developed in particular from the work of Bruno Latour and his thesis that modernity involved the artificial separation of the natural and social realms. Latour contends that, in fact, modernity was founded on heterogeneous processes in which humans became increasingly bound to non-human, such that the world is composed of hybrid assemblages of human and non-human actants (Latour, 1993; see also Murdoch, 2003; Whatmore, 2002).

Murdoch (2003) identified a succession of rural networks and relations that exist in hybrid form. These include various networks of agricultural production that combine human, living non-human and technological components; processes of development that transform landscapes for social or economic purposes; recreational activities dependent both on complex technologies and on particular landscapes and climates; and patterns of transportation comprising new human–machine relations (see also Murdoch, 2006). He also noted that the rural is governed through hybrid networks, highlighting Enticott's (2001) study of the network involved in decision-making about the alleged link between badgers and bovine tuberculosis in Britain, which comprised human and non-human participants.

Murdoch further contended that the rural itself is hybrid. The potential for understanding rural localities as hybrid assemblages is examined by Rudy (2005) in a study of Imperial Valley in southern California. Rudy critiques conventional representations of Imperial Valley as a watershed and as a region, which he notes reproduce the separation of the natural and the social. Instead, he proposes an understanding of Imperial Valley as a 'cyborg' (a 'coevolved, hybrid, and cybernetic organism' (p. 28)), borrowing the terminology from Haraway (1991). As a cyborg, Imperial Valley is understood as comprised by diverse elements including Colorado River water; migratory waterfowl; the intentionally seeded food chain of the Salton Sea; the San Andreas Fault; Mexican field labour; public universities' extension services; global markets and supply chains; international biotechnology, chemical and seed conglomerates; and state and federal regulation of water rights, regulations and markets. These elements are bound in numerous complex inter-relations, such that 'the Valley maintains connection with multiple bodies and categories of phenomena historically associated with the natural, the social and/or the technoscientific while never falling completely within any one of those categories' (Rudy, 2005: 29). The co-constitution of Imperial Valley means that individual social, economic or environmental problems cannot be considered in isolation from other elements of the network, so, for example, 'the ecological problems of the Valley cannot be held distinct from those of labor struggles and worker health, much less from (agri)cultural practices and community (re-)development' (p. 30).

Recognizing the rural as hybrid hence requires new ways of knowing the rural, exposing the folds and crevices of rural space that remain hidden to perspectives focused only on the social, or only on the natural. As Murdoch (2003) observed:

> The countryside is hybrid. To say this is to emphasize that it is defined by networks in which heterogeneous entities are aligned in a variety of ways. It is also to propose that these networks give rise to slightly

> different countrysides: there is no single vantage point from which the whole panoply of rural or countryside relations can be seen.
> (Murdoch, 2003: 274)

Relational approaches can help to illuminate this 'topologically complex' rural space, but Murdoch acknowledged that relational perspectives can themselves be limited in scope. In particular, he suggested that there are times when human actors need to be considered as more than just equal participants in a hybrid network, that 'while humans are clearly enmeshed in networks of heterogeneous relations, they also retain distinctive qualities as members of such networks and these qualities can be seen, at times, as "driving" sets of changes in the socio-natural world' (p. 275). As such, Murdoch proposes that the relational approach should not be seen as supplanting social constructivism and exploration of engagements with rural natures, but that it should be employed alongside these established approaches as an additional tool in the kit of the rural geographer.

Rethinking urban–rural relations

The relational approach can also help us to rethink the relationship between the rural and the urban. The rural has always been defined and imagined in relational terms, as relative to urban space and society. At the same time, actual social and economic relations – the networks and flows of people and goods, capital and power – have always transgressed the discursive divide of urban and rural. Functional and political-economy conceptualizations of rurality have struggled to resolve this paradox, leading to the suggestion that the 'rural' should be abandoned as an analytical term (Hoggart, 1990); whilst social constructionist approaches have focused on the discursive realm to the neglect of actually existing social and economic relations. The relational approach, in contrast, permits us to recognize the diverse networks and flows that criss-cross rural and urban space and the hybrid forms that result as being part of the very constitution of both the rural and the urban.

The mixing of urban and rural has been most commonly observed with respect to the perceived 'urbanization' of the rural. Urbanization involves a number of different processes, all of which have been represented as threatening the 'rural' essence of the countryside, yet which, tellingly, all depend on a discursive classification of urban and rural forms fixed at some unspecified point in history. Thus, urbanization has been associated with the expansion of non-agricultural economic activities, and with the orientation of rural production to serve urban markets. Yet, as Barnett (1998) observed, urban economic dominance has a long history: 'the countryside is in urban hands already, as it has been since the city generated its trade and capital' (p. 342). Less tangibly, urbanization is also identified with the permeation of urban cultural practices, attitudes and consumption patterns. Yet here, urbanization is often conflated with globalization, such that any external cultural influence threatening to displace established rural beliefs and traditions is characterized as 'urban' (Cloke, 2006).

Assertions about the urbanization of the countryside are hence based on subjective characterizations of urban and rural. They do, however, point to the empirically demonstrable fact that rural localities today are tied into networks centred on urban sites of economic, political and cultural power, and this extends to the discursive representation of the rural. As Svendsen (2004) has observed in Denmark, 'all contemporary Danish rural identities can be seen as embedded in – or, at least, influenced by – urban terminology and practice' (p. 89).

Less commonly remarked on is the 'ruralization' of urban space and society. Again, this phenomenon is composed of several different elements. First, there is the presence of rural populations in the city. Historically, the rural elite commonly maintained urban properties and participated in urban social and political affairs, such that rural power was always in part dependent on networks embedded in urban settings, and urban power on networks embedded in rural settings (Woods, 2005b). At the other end of the social order, rural migrants to cities have always taken with them traditions and practices from their home regions, producing hybrid cultures that remain significant in many cities of the global south, as will be discussed further in Chapter 6.

Second, there is the incorporation of rural landscape features into the urban built environment, usually informed by a moral geography that contrasted the purity and orderliness of the countryside with the chaos and degeneration of the city. Examples include the 'garden cities' and 'garden suburbs' pioneered by figures such as Ebenezer Howard (Bunce, 1994), as well as more contemporary attempts to replicate 'village' life in urban gated communities and the appropriation of rural iconography in urban commercial developments (Wilson, 1992). City parks also represent, as Jones and Wills (2005) observe, efforts to 'bring the country to the metropolis'. The first urban parks were hunting grounds for the elite, whilst their later transformation into public pleasure grounds involved reproducing the parkland landscapes of country estates. Central Park in New York, for example, originally boasted not only formal gardens, carriage drives and areas of woodland, but also a resident flock of sheep grazing in meadows.

Third, if agriculture is defined as rural, then sites of urban agriculture can be positioned as pockets of rurality within the city. Urban agriculture is important in many cities in the global south, where it makes a vital contribution to local food supply. Examples of urban agriculture can range from backyard chicken-runs and rooftop horticulture to the use of road verges and waste land for grazing to commercial farms (Lynch, 2005). In some countries, urban agriculture has been officially encouraged, such as in Nigeria where vacant public lands in urban areas are available for cultivation without charge (Potter *et al.*, 2008), whereas in other cities it is regarded as posing health and sanitation challenges (Lynch, 2005). In the global north, urban agriculture is regaining a presence in many cities having been suppressed by planning and sanitation regulations. City farms were established in cities such as London and Bristol in the 1970s with primarily educational objectives (Jones, 1997), but have expanded with growing interest in alternative food networks, alongside the increasing popularity of urban allotments in the UK, and the use of community gardens for food production in cities such as New York and San Francisco (Allen *et al.*, 2003). The engagement of urban residents in community-supported agriculture schemes linked to peri-urban farms, and the establishment of inner-city farmers' markets (Allen *et al.*, 2003; Jarosz, 2008), can be seen as part of the 'ruralization' of aspects of urban life.

Fourth, urban consumption cultures can exhibit preferences for commodities associated with rural iconography. Examples include periodic fashion for 'rural chic', as expressed though clothing and interior design, and the popularity with urban residents of sports utility vehicles (SUVs) originally designed for off-road rural travel (colloquially known in Britain as 'Chelsea tractors' after a prosperous London neighbourhood). The association of these consumption preferences with the affluent 'service class' of professionals commonly identified as a key group in counterurbanization processes (Thrift, 1989), suggests that they are driven in part by aspirations towards a rural lifestyle, or at least by a desire to advertize that the owner can transgress the urban–rural divide by escaping to the country for recreation.

Fifth, Urbain (2002) describes the expansion of the urban field into the countryside as a process of ruralizing the city as much as an urbanization of the rural, as Cloke summarizes:

> Urbain insists that the spread of the city out into the country has effectively ruralized a very significant part of the urban. Given that the nature of the city has been radically changed, both by centralizing tendencies and by decentralizing practices, it can be argued that an important slice of contemporary urbanity can now be found in the village, and that the urban form thereby now encapsulates very strong rural characteristics and influences.
>
> (Cloke, 2006: 19)

This last assertion about the ruralization of the urban directs our attention to the liminal zones where the entanglement of the rural and the urban is most pronounced. These include peri-urban localities that are characterized by 'rural' landscapes and settlement forms, but which are locked into urban labour markets and service centres (see Bertrand and Kreibich, 2006; Hoggart, 2005; Masuda and Garvin, 2008), and 'exurban' communities in North America.

The concept of 'exurbia' in particular resonates with Urbain's thesis, as it differentiates the process of exurbanization from both suburbanization and the colonization of established rural communities by urban migrants. Exurbanization, it is suggested, 'involves scattered, isolated pockets of residential development some distance from an urban center in areas possessing high aesthetic values and natural amenities' (Larsen *et al.*, 2007: 421). It is associated with the subdivision of ranches and farmland for medium-density development, which frequently takes the form of 'leapfrog' or 'checkerboard' development, featuring ranch houses, 'pods' and 'pork chop lots' (Hayden, 2004), such that housing is interspersed with open land, but still approaches urban density levels.

Exurban development is also often speculative, driven by property developers who 'invoke images of an idyllic rural lifestyle to market stability, timelessness, and tranquillity to middle-class consumers' (Larsen *et al.*, 2007: 424). Disjunctures between the rural idyll branding of exurban developments and the lived experience of a suburban-type built environment, together with concerns at the social and environmental impacts of rapid development, have helped to make exurbia a contested space. Larsen *et al.* (2007), for example, describe residents' concerns about exurban development threatening strong local attachments to place in Garden Park, Colorado; whilst Walker and Fortmann (2003) detail conflict over the incorporation of environmental principles into development planning in Nevada County, California.

Lacour and Puissant (2007) characterize the hybrid spatial forms and practices that emerge from the entanglement of the rural and the urban as a condition of 're-urbanity'. This they see as the re-articulation of urbanity through rural localities, the renaissance of which is produced by an iterative fusing of elements of rural and urban attractiveness. Urban migrants are attracted to rural localities by ideas of the rural idyll, including closeness to nature, solidarity and community spirit, but rural localities are made acceptable to urban migrants by their endowment with urban features such as high-quality public services, artistic and cultural activities, and cultural diversity. Vibrant rural communities give rise to forms of creative expression commonly identified with urbanity, and produce new commodities that are dependent on urban consumption. In turn, ideas of solidarity, community spirit and identity associated with rural life are applied in urban development in an attempt to match the revitalization of rural localities by regenerating areas of the city. In this way, Lacour and Puissant position us in a time of transition and rebirth that is marked by both the urbanization of the rural and the ruralization of the urban.

CONCLUSION

The purpose of this chapter has been to examine the ways in which the rural has been imagined into being over time, and the ways in which we come to know the rural, both as individuals seeking to make sense of the world around us and as geographers interested in studying the enduring power of the idea of the rural to shape places and spaces. The chapter approached these questions from three perspectives. First, it traced the development of the rural as an idea from the earliest recorded writing of Theocritus and Virgil, through the mediaeval period to the transportation of European ideas of rurality around the world in the era of colonization. As was discussed, settlers attempted to shape the rural spaces of the rest of the world to European ideas of rurality, but these ambitions collided with the physical realities of colonial territories, producing new hybrid rurals which in turn impacted back on European ideas. One example of this was the tension that had existed in mediaeval European imaginings of the rural between the pastoral idyll and the anti-idyll of the wilderness. This dichotomy initially fed the drive to conquer and tame the 'wilderness' encountered in the Americas, Africa and Australia, yet over time North Americans in particular came to value the rural as wilderness and protect wilderness spaces. The subsequent global dissemination of this idea coincided with growing concerns in Europe about urban encroachment on the countryside, producing a preservationist movement that translated the imagined divide of the urban and the rural into a regulated spatial order.

Second, the chapter explored the epistemological question of how we know the rural. It revealed both the partial nature of various discourses of rurality and their power to have material effects in rural space. Quantitative discourses, for example, are problematic in defining the limits of rural space, yet the lines that they do draw have real effects on policy implementation and funding programmes. Media discourses often exaggerate stereotypes of rural life, but they also shape public expectations of the rural. Lay discourses of the rural are situated knowledge grounded in particular experiences of rural life, but inform the way in which rural residents engage with processes of change. Third, the chapter engaged with recent work in rural geography that has adopted a relational approach to the rural, reimagining the rural as a hybrid space constituted by networks and flows involving both human and non-human actants. In particular, it was suggested that the relational approach presents opportunities for rethinking the relation of the urban and the rural, recognizing both the urbanization of the rural and the ruralization of the urban.

Each of these perspectives reveals part of what is understood by rurality, illuminating particular aspects of Halfacree's three-fold model of rural space (see Chapter 1). Attempts by governments to define and map an official classification of rural space, for example, are part of the formal representation of the rural, as are various policy, media and academic discourses of rurality. Lay discourses of rurality, meanwhile, articulate the everyday lives of the rural, which are also performed in non-representational ways (see also Chapter 7), and which are 'inevitably incoherent and fractured' (Halfacree, 2006: 51), giving rise to multiple, disjointed and hybrid rural experiences. Both of these types of discursive representation engage with the materialities of rural localities, inscribed with distinctive spatial practices of production or consumption activities (discussed further in Chapters 3 and 4, respectively), which in turn are located within wider relational networks.

The rural is not a unified, discrete and unambiguous space. Rurality is always a complex and contested category, and places can only be described as being rural insofar as they exhibit more characteristics of rurality relative to other alternative spatialities. Thus, at least some characteristics of urban space can be found in most 'rural' places, but elements of rural space can also be found in towns and cities. The point here is that arguments that suggest that urban space has become all-pervasive, or that the rural is no longer a meaningful category, are mistaken. Rurality continues to be a powerful cultural concept, as it has been throughout history, and continues to have a material effect in shaping the social, economic and political geographies of large parts of the world. The following chapters examine in more detail some of the ways in which the imagined rural is grounded and enacted and has material effects.

FURTHER READING

Several of the themes touched on in this chapter are discussed further in contributions to the *Handbook of Rural Studies*, edited by Paul Cloke, Terry Marsden and Patrick Mooney (2006). These include the chapter by Paul Cloke on 'Conceptualizing rurality' which surveys different ways of defining and approaching the rural; Brian Short's chapter on 'Idyllic ruralities' which traces the evolution of the idea of the rural idyll; and Jonathan Murdoch's chapter on 'Networking rurality'. The historical development of the idea of the rural in Britain and North America is comprehensively recorded by Michael Bunce in *The Countryside Ideal* (1994), whilst Jill Casid explores the cultivation of the rural landscape as strategy of colonization in *Sowing Empire* (2005). The role of television dramas in reproducing media discourses of rurality is further examined by Martin Phillips, Rob Fish and Jenny Agg (2001) in the *Journal of Rural Studies*, whilst the concept of lay discourses of rurality is developed by Owain Jones (1995), also in the *Journal of Rural Studies*. The best introduction to relational perspectives on the rural is provided by Jonathan Murdoch's chapter in *Country Visions*, edited by Paul Cloke (2003). Alan Rudy's paper on Imperial Valley in the *Journal of Rural Studies* (2005) is an accessible case study of a relational approach applied to a rural locality.

3

EXPLOITING THE RURAL

INTRODUCTION

What is the countryside for? This is a question that it appears is being asked with increasing frequency in conferences and seminars and press articles, as contemporary society attempts to come to terms with the consequences of rural restructuring. Yet, historically, the answer to this question has been simple. Throughout history the primary function of rural space has consistently been understood as the supply of food and natural resources, including minerals, fuel and building materials. Whereas the town and the city have been constructed as centres of trade, manufacturing, cultural exchange, social provision and political administration, the countryside has been associated with the exploitation of its natural resources through farming, forestry, mining, quarrying, fishing, hunting and energy production. The idea of the rural as a space of production and exploitation has arguably been the single most influential idea in shaping rural space, giving rise to particular landscapes, settlement patterns, forms of social organization, political structures and policies, and economic systems.

Originally, the resources of rural land were exploited for domestic survival – the practice of subsistence agriculture that remains important in some regions of the global south (Waters, 2007). The development of towns and cities created a demand for food and other resources that initiated trade in surplus agricultural produce, as well as the opening of quarries and mines to obtain mineral resources. Over time, commercial farming, forestry, mining and quarrying were developed expressly to service market demands, and collectively came to form the mainstay of the rural economy. A further progression occurs with the development of agrarian capitalism, or resource capitalism (to include mining, quarrying etc.), in which production was no longer driven by demand, but rather by the imperative to make a profit, such that producers started to search out new markets and find new ways of increasing the return on their capital investment.

Capitalist modes of production have dominated the exploitation of rural resources in the modern era, but the conditions of operation of agrarian and resource capitalism have varied both temporally and geographically, framed by different political-economic regimes. This chapter hence examines the capitalist exploitation of rural resources under three different political-economic regimes. First, it analyses the development of an unrestrained resource capitalism in late nineteenth-century California, drawing particularly on the work of George Henderson and Richard Walker. Although the story of Californian resource capitalism is exceptional, the case study is used to identify a number of features of resource capitalism that have more generic expression. This is illustrated by a brief discussion of forestry in the Pacific Northwest, plantation capitalism in the tropics and American south, and the expansion of the resource frontier for oil and coal extraction in northern Canada.

Second, the chapter examines the rise of state-sponsored productivist agriculture in the post-war period, with a particular focus on Europe, but also comparing the European experience with the 'Green Revolution' in the global south. It will be argued that whilst productivism prioritized agricultural production as the primary goal of the rural economy, it also recognized that agricultural production occurred within a broader socio-economic system and incorporated social goals that were missing from pure resource capitalism. Yet, the discussion also exposes the paradox of productivism in that the reforms it introduced had the effect of reducing the dependency of the rural population on agriculture, thus opening a space for alternative discourses of the rural, which eventually challenge the supremacy of productivism and initiate debates around the post-productivist transition and multifunctionality.

Third, the chapter considers the emergence of 'multifunctional agricultural regimes', in which the functions of agriculture – and of rural space – are recognized as being more than to maximize the production of food and fibre. It is argued that multifunctionality is a more useful concept than the post-productivist transition, as it acknowledges that productivism has not only continued to shape certain rural localities, but has also become more advanced in the form of 'super-productivism'. At the same time, multifunctional regimes also permit alternative models of production, such as organic farming, yet, as is discussed, even these sectors are subject to the accumulation imperative of agrarian capitalism.

In each section, the discussion seeks to show how the modes of production concerned are associated with particular discursive representations of the rural economy, and how these in turn are translated into material practices with effects on the landscape, society and economy of rural localities. As such, the chapter is focused on the connections between two facets of Halfacree's three-fold model of rural space (see Chapter 1). Primary production – of food, fibre, timber, fuel and minerals – is commonly perceived to be one of the 'distinctive spatial practices' through which rural localities are inscribed; but the conditions for these activities are framed by formal representations of the rural that encode the rural as a space of production. The production-based economy also intersects with the third element of Halfacree's model, the everyday lives of the rural, but that dimension is examined in more detail in Chapter 7, informed by post-structuralist perspectives on performance and more-than-representational geographies. This chapter, in contrast, adopts a more structuralist approach, concentrating on the structures that underpin the rural economy, and which locate it within the broader capitalist political economy (see Cloke, 1989a for more on the political economy perspective in rural geography).

RESOURCE CAPITALISM AND THE RURAL ECONOMY

Resource capitalism and rural realism in California

In 1840, California stood on the edge of the European-American world. It was part of the Mexican republic, its non-indigenous population of less than 15,000 people concentrated into a scattering of mission and trading settlements along the coast. The interior landscape of desert, mountain, forest, wetland and scrub was home to around 70 native tribes and large cattle *rancheros*, producing cowhide and tallow for trade with east coast merchants. The discovery of gold at Coloma in 1848 changed everything, prompting a 'gold rush' that brought an estimated 300,000 migrants to the territory. In 1850, California was admitted to the United States, and in the ensuing decades it experienced an economic boom that was founded on an unprecedented exploitation of natural resources. As Walker (2001) observes, 'wave after wave of resource accumulation figured in the state's rapid growth: gold, silver, wheat, citrus, timber, copper, hydropower, petroleum, sardines' (p. 172).

There are many rural regions that have experienced resource booms. One of the exceptional features of California's trajectory is that it so rapidly achieved pre-eminence across such a diverse range of commodities. It was the world's most important mining district at the close of the nineteenth century, responsible for one-fifth of global gold production; it became the top agricultural producing state in the US in 1925, and still leads production in 76 crops; it was one of the top three timber-producing states during the first half of the twentieth century; and was the leading oil producer in the world between 1905 and 1930 (Walker, 2001, 2005). The exponential development of these industries during the late nineteenth and early twentieth centuries transformed the economy, society and landscape of California, in effect manufacturing the modern Californian countryside.

However, as Walker (2001) comments, 'California's resource bonanzas were not passive withdrawals form the earth' (p. 169). They were rather the creation of 'resource capitalism': a mode of production designed to maximize the financial return on the exploitation of the land for its natural resources. Resource capitalism requires efficient mechanisms for extracting value from nature at minimal cost, and as such is characterized by David and Wright (1997) as involving 'intensity of search; new technologies of extraction, refining and utilization; market development and transportation investments' (p. 171). These principles apply across the different primary industries and produce a dynamic that demands continual innovation and improvement. At the same time, Henderson (1998) notes that the fact that farming is centred on nature is crucial, recognizing nature as an active agent that can act to limit or constrain capitalist accumulation (the same observation applies to mining and other extractive industries). Thus, whilst the primary narrative of rural California is one of phenomenal growth, the story is punctuated with episodes of crop failures, mine collapses, floods and other events where nature intervenes in unscripted ways.

Resource capitalism was already established in different variations in many places around the world by the time of California's resource boom. What made California different was the specific combination of an abundance of natural resources of various kinds with very few pre-existing socio-economic structures to constrain the application of capitalist principles. Resource capitalism was hence able to develop in California in something close to its purest form. Walker (2005) notes that unlike farming in Europe or the eastern United States, Californian agriculture 'has always been for profit' (p. 79), and that it 'has only become more thoroughly capitalist with time' (p. 13). The full-bodied development of resource capitalism in California was even noted by Karl Marx, who commented to a correspondent that 'California is very important ... because nowhere else has the upheaval most shamelessly caused by capitalist centralization taken place at such speed' (quoted in Walker, 2001: 182).

The discursive framing of Californian resource capitalism starts from the representation of the territory as a land of opportunity and untold riches, a promise that inspires the first wave of gold rush in-migrants. As the exploitation of rural California develops, the discourse of opportunity is married with an idyllic representation of the state as a place of fruitfulness and good climate in a bid to attract the additional settlers required to feed the dynamic of the resource boom. Walker (2005), for instance, reproduces a poster issued by the California Immigration Commission in 1885 which advertises the 'Cornucopia of the World', with room for millions of immigrants, millions of acres of land untaken for a million farmers, and a 'climate for health and wealth, without cyclones or blizzards' (p. 21). A new genre of literature, which Henderson (1998) labels 'rural realism', documented the settlement and exploitation of the land, eschewing an idyllic representation of the 'natural' countryside for a celebration of the heroic role of capital in manufacturing the rural:

> Rural realism appropriated stock images – of fruited plains, embowered farmsteads, glistening rivulets – only to better assert that the 'rural' in rural realism would be no refuge from capital but would be one of the most desired places for it ... More than once are the characters who bring rural realism to life led to the fields and orange

> groves *by* bankers and developers rather than running there to get away from them. Capital could *bring the rural into being and, recursively, would be the better for doing so.*
>
> (Henderson, 1998: xvii, italicization in the original)

The contemporary discursive framing of the resource economy hence also prescribed the social relations required to translate the dream of opportunity into material reality. Central to these was the principle of private property as the foundation for resource extraction and production. The nascent state government facilitated this development by sponsoring the dispossession of native tribes, disregarding tactics of squatting and intimidation used to wrestle land from rancheros, and enshrining a radically populist system of mining claims in legislation (Walker, 2001, 2005). In effect, it wiped the map clean for colonization by small-scale prospectors, farmers and ranchers who provided the entrepreneurial energy for exploiting the land.

Small-scale operators required capital to develop their enterprises and the second key element of the Californian rural political economy was that the profits of resource exploitation were retained and reinvested in the state. A banking system was established on mineral profits, which it reinvested in agricultural development (Walker, 2001, 2005). As Henderson (1998) details, the circulation of financial capital was critical to the development of Californian agriculture, with the extension of credit enabling farmers to overcome the impediments presented by nature to the smooth accumulation of capital through agricultural production (the investment needed to prepare the land; the risk of drought, flood, crop failure, etc.). Moreover, the credit system also involved a further form of discursive representation of the rural, as rural landholdings were converted into 'fictitious capital' based on their potential productive value as the basis of security for loans.

The exploitation of Californian nature by resource capitalism brought into being the rural landscape and rural society of modern California. It is important to note, though, that these landscape and social effects were not by-products of resource exploitation, but rather were integral to strategies of capital accumulation. The landscape was transformed, for example, by land reclamation projects, such as the draining and clearing of the Sacramento – San Joaquin Delta for arable production, and irrigation schemes, which had created 5,000 miles of irrigation canals by 1900

(Henderson, 1998; Walker, 2005). More subtly, the landscape was populated by imported crops and livestock – sheep, cattle, mules; peach, apple, pear, apricot, orange and cherry trees; vines, strawberries, olives, figs, artichokes and rice, among many others – and later by the modified breeds and varieties propagated by the state's own agricultural scientists (Walker, 2005). Radically, these crops and livestock were raised by specialist producers, with Californian agriculture based on monoculture from the start, and regional specialization produced a patchwork landscape with localities perfecting the cultivation of specialty crops (*Ibid*).

The new rural landscape was also characterized by the network of railways that developed to export the produce of mines, forests and farms, and by the town sites established by the railroad, mining, timber and cattle companies to service the resource economy (Walker, 2001). In contrast to Europe, or even longer-settled parts of North America, there was no existing rural population in California to provide the labour force for resource capitalism. Initially, migrants were attracted from elsewhere in the United States by comparatively high wages. However, as the dynamics of capital accumulation demanded mass, low-cost and trouble-free labour, Californian agriculture turned to hired immigrant workers, drawn in waves from Mexico, China, Japan, Vietnam and India. As Walker (2005) remarks, 'the most striking thing about the labor system in California farming ... is not just that it involves hired labor but that it has used one group after another, in a vast, repetitive cycle of recruitment, employment, exploitation, and expulsion' (p. 66) (see also Box 3.1). Immigrant labour serviced resource capitalism, but was encountered by rampant racism, which was manifest spatially as well as socially with the establishment of separate migrant worker villages (González, 1994; see also Henderson, 1998; Mitchell, 1996; Walker, 2005). Most notorious was the *bracero* programme, which brought in Mexican workers as a form of indentured agricultural labour: 'contracted by government, housed in closed camps, bused to the fields, and sent home to Mexico once the season was over' (Walker, 2005: 72).

Finally, the Californian countryside created by resource capitalism was locked from the start in a symbiotic relationship with the state's urban centres. Cities such as Los Angeles and San Francisco grew on the wealth generated by resource exploitation, including the development of derivative industries such as food processing, oil refining and the manufacturing of farm machinery. The same cities also formed the first markets for agricultural produce and as the pursuit of new markets intensified, it was

Box 3.1 THE RURAL LABOURING BODY

Labour is a fundamental input for resource capitalism, especially for the type of intensive arable farming and horticulture that developed in California in the late nineteenth century. Labour was required to prepare the land, plant seeds, tend the fields, harvest crops and pack produce. In common with other inputs, the imperatives of capital accumulation demanded that labour was cheap, reliable and pliable. Yet, as Mitchell (2000) observes, 'cheap labour is no natural commodity; it needs to be *made*, be conditioned and served up to growers when and where they need it, and, ideally, for no longer than they need it' (p. 244). Moreover, Henderson (1998) argues that labour differs from other agricultural inputs in that its use is constrained by the physical limitations of the human body:

> legions of bodies will tramp the ground that feeds the roots; they will temporarily interrupt sunlight as they lean over and work their fingers through stems or vines to find ripe berries, harvest grapes, or cut asparagus. Sometime during the heat of the day, these legions will pause for some food and drink. A portion of agrarian capital will come to a halt.
>
> (Henderson, 1998: 81)

The ideal rural labouring body therefore needed to be moulded both to maximize the value of its input and to minimize the disruption inevitably caused to the process of capital accumulation. California's agrarian capitalists initially hoped to mould the bodies of white American migrant workers to create 'a much more efficient, much less militant, house for labor power' (Mitchell, 1996: 88). As these efforts were frustrated by the unionization of the agricultural workforce, growers increasingly turned to foreign labour. As Mitchell (1996) records, the employment of migrant workers from Mexico and East Asia was justified by racist representations of the bodies of the immigrants as being 'naturally suited' for farm work, in two ways. First, the physical stature of migrant worker bodies

continued

was represented as being 'naturally' appropriate for farm work, notably the short stature of Chinese workers. Second, as non-whites were represented as being more primitive than whites, it was accepted that their bodies could be disciplined and controlled in ways that would be unacceptable if applied to white Americans.

Thus, Mitchell reports the explanation of one Californian farmer, recorded in archives, that 'We want Mexicans because we can treat them as we cannot treat any other living men ... We can control them at night behind bolted gates, within a stockade eight feet high, surmounted by barbed wire ... We make them work under armed guard in the fields' (Mitchell, 1996: 88). Similarly, he quotes another contemporary report of farms making 'imported Mexicans work for ten or twelve hours a day, handcuffing them at night to prevent their escape' (*Ibid*: 90).

Mitchell focuses on the primitive and unsanitary conditions in which migrant farmworkers were forced to live and the restrictions imposed on the mobility of their bodies. In the fields, meanwhile, the bodies of farmworkers were disciplined by long working hours and punishing schedules, the incentives of piecework, poor equipment, limited safety precautions, and exposure to heat and pesticides – the latter especially after the Second World War (Nash, 2004).

Although working conditions have improved in the intervening decades, the bodies of migrant agricultural labourers in California (and elsewhere) continue to be moulded and abused in the interests of capital accumulation. Mitchell (2000) considers the bodily practices that are part of strawberry cultivation in present-day California, noting that 'strawberry picking and plant maintenance requires that workers spend the day doubled over at the waist as they work their way down a row, often standing or stretching only when they reach the end' (p. 236). These conditions, he observes, marks the bodies of the workers: 'Back injuries are exceedingly common. So too are respiratory ailments from inhaling pesticides and dust, severed fingers and hands, and progressively developed allergic reactions to strawberry juice, flowers and leaves' (*Ibid*).

As such, the physical demands on labour for capitalist agriculture mould and mark the bodies of farmworkers. Mitchell describes this as a form of violence against the body, thus arguing that 'California agriculture is structured through violence, the violence done to bodies through incessant stoop-labor, pesticide exposure, and horrible living conditions, as well as the violence endemic at the border, in their home villages, in the campaigns to organize farmworkers (and resistance to these by growers), and in the cities and towns that farmworkers live in up and down the agricultural valleys of the state' (Mitchell, 2000: 238).

Further reading: Mitchell (2000).

urban-based conglomerates that led the way in innovations such as supermarkets, canned food and new food commodities to sell Californian produce (Walker, 2005). As new markets developed, rural California was rapidly integrated into global networks and flows. This is strikingly illustrated by Henderson (1998), who quotes a character in Frank Norris's 1901 novel, *The Octopus*, explaining that:

> The most significant object ... was the ticker ... The offices of the ranches were thus connected by wire with San Francisco, and through that city with Minneapolis, Duluth, Chicago, New York, and ... most important of all, with Liverpool. Fluctuations in the price of the world's crop during and after the harvest thrilled straight to the office of Los Muertos, ... The ranch became merely the part of an enormous whole, a unit in the vast agglomeration of wheat land the whole world round, feeling the effects of causes thousands of miles distant – a drought on the prairies of Dakota, a rain on the plains in India, a frost on the Russian steppes, a hot wind on the llanos of the Argentine.
>
> (quoted in Henderson, 1998: 141)

Variations on resource capitalism

The extended discussion of Californian resource capitalism above has provided a detailed example of how the framing of the rural as a space whose primary function is the capitalist exploitation of natural resources has material effects in shaping the economic structure, society and landscape of rural areas. Whilst California is an extreme and unique case, resource capitalism and agrarian capitalism are practised in various forms across the world, replicating many of the features observed in California. Indeed, it is important to emphasize that neither resource capitalism nor agrarian capitalism were invented in California. Early expressions of agrarian capitalism have been identified in England and the Netherlands as far back as the fifteenth century, and agrarian capitalism had gradually spread throughout England by the early nineteenth century (Shaw Taylor, 2005; Whittle, 2000). This transformation created the distinctive enclosed landscape of rural England, as well as its social and economic structure. In the United States, agrarian capitalism flourished on the Great Plains just as surely as on the west coast, but in the Mid-West it was characterized by the incorporation of pioneer subsistence farmers into the marketplace and by the dominance achieved by conglomerates such as ConAgra that provided the infrastructure for capitalist accumulation, including grain mills and food processing.

Also well entrenched by the nineteenth century was plantation capitalism, a tool of colonialism, with the terms 'plantation' and 'colony' even being used interchangeably (Casid, 2005). The plantation system was developed in the Canary Islands, Cape Verde Islands and island of São Tomé off west Africa in the fifteenth and sixteenth centuries, and exported by Spanish and Portuguese colonists to the Caribbean and South America (Potter *et al.*, 2008). Plantations represented the most rapacious form of European exploitation of the rural lands of Africa, Asia and America, designed solely for the production of cash crops to feed the cravings of European consumers and generate wealth for European merchants. As Casid (2005) records, the establishment of plantations required the scouring of the pre-existing landscape, involving 'vast deforestation, the clearing of the undergrowth and the burning of any remaining roots' (p. 7). This process made rural land empty, and thus legally available for appropriation by colonists who replanted with crops introduced from around the world for their commercial potential:

> Not only were the main cash crops of the plantation system – sugarcane, coffee, and indigo – transplants, but plant transfers to the Caribbean from Europe, Asia, Africa and the South Pacific so radically

> transformed the landscapes of the Caribbean islands that those species of flora most symbolically associated with the 'tropics' were precisely those plants by which the British grafted one idea of island paradise onto another. Bamboo, logwood, cashew, casuarina, royal palm, imortelle, coconut palm, citrus, mango, tamarind, breadfruit, banana, bougainvillea, hibiscus, oleander, poinsettia, thunbergia, and even pasture grass (guinea grass from West Africa) were all colonial transplants.
>
> (Casid, 2005: 7)

Plantations were sites of intensive industrial agriculture, fine-tuned to maximizing capital returns through economies of scale in production and processing. In time, however, the advantages of plantations over peasant agriculture are eroded by the limitations of crop and task specialization and the inefficiencies involved in managing labour over large areas, especially in regions such as South East Asia where competition for land intensified (Hayami, 2004).

Restraining labour costs was critical to plantations, and in the Americas, in particular, slave labour was a key element in the plantation mode of production. Between the sixteenth and nineteenth centuries nearly 15 million individuals were forcibly transported from Africa to the Americas to work as slaves, mostly on plantations (Potter *et al.*, 2008). However, as Hong (2001) describes for the sugar plantations of the southern United States, plantation capitalism was not a pure slave mode of production, but 'was framed not only by a semifeudal class structure and coerced labor system but also by a capitalist imperative of accumulation and commodification' (p. 23). Capital, land and technology were also fundamental inputs, and numerous management strategies were innovated and applied to perfect the exploitation of slave labour. In particular, time was critical for the regulation of human and non-human components of the plantation production system, such that Hong argues that 'it was plantation capitalism that placed rural societies on the solid fundament of modern time consciousness' (p. 3). The cultural conditions and practices fashioned in plantations permeated through local society, and long after the abolition of slavery, the legacy of plantation capitalism persists in the landscape, culture, politics and social and economic structures of the rural regions that it once dominated.

A further variation in the model of resource capitalism can be found in regions historically dominated by commercial forestry, such as the 'Douglas fir country' of the Pacific Northwest of America. As a capitalist enterprise, forestry shares many of the challenges experienced by agriculture, including exposure to the unpredictability of nature and the time delay involved in obtaining returns from capital investment, thus requiring the translation of rural land and resources into 'fictitious capital' (Henderson, 1998; Prudham, 2005). Indeed, the time-scales involved in commercial forestry dwarf those of agriculture. Douglas fir trees in the Pacific Northwest, for example, take between 50 and 80 years to reach maturity for commercial harvesting (Plate 3.1). Consequently, forestry is a land-hungry industry, Prudham (2005) observing that the typical sawmill in Oregon requires access to around 100,000 acres of timber plantation to operate over the long term.

The particular demands of the capitalist mode of production in commercial forestry and the lumber industry have produced distinctive forms of economic, social and environmental relations in the rural regions that they dominate, as Scott Prudham (2005) demonstrates in his study of Oregon. The imperative to maximize returns on capital investment by

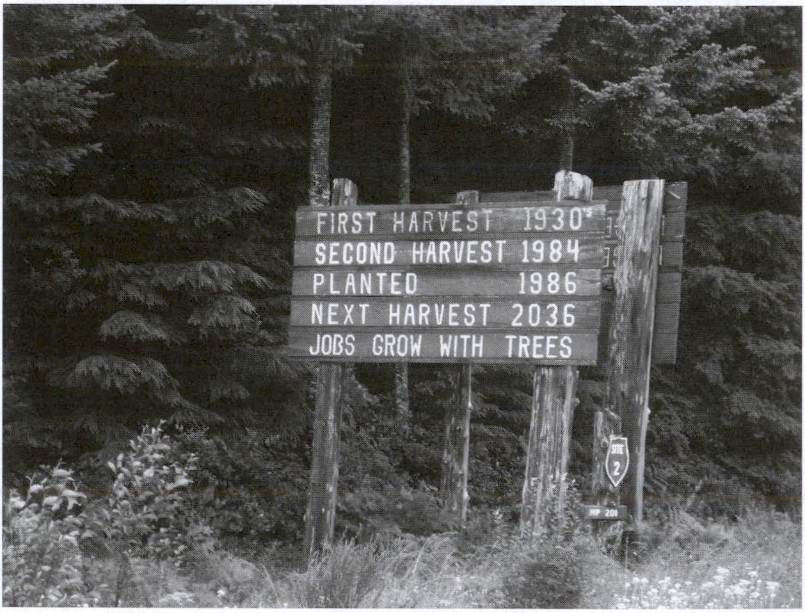

Plate 3.1 Time-lag in commercial forestry on the Olympic Peninsula, Washington State (Photo: author)

minimizing the waste from harvested trees, for example, promoted the development of processing industries specializing in producing commodities that consume wood in different sizes and in different forms, from whole log conversion as lumber, to plywood manufacture, pulping and paper-making. Similarly, the costs involved in transporting cut timber to sawmills enforced a spatial fixidity that encouraged sawmills and processing plants to be developed in close geographical proximity to timberlands, which in turn mitigated against the achievement of economies of scale in large plants. As such, wood processing was historically an industry characterized by small-scale operations. The expansion of the wood industry in Oregon and Washington hence saw the number of sawmills in the two states increase from 36 in 1850 to 1,243 in 1920, most of which primarily served local markets and had small workforces. Corporate expansion instead came through the consolidation of forest land holdings, with Prudham (2005) reporting that by 1913, the 16 largest timberland owners in the Pacific Northwest controlled more than 40 per cent of private forest land in the region.

Forestry and wood processing both required the input of labour, but the industry also needed to protect land for tree cultivation, thus constraining urban development. These factors led to a distinctive settlement pattern of small towns dominated socially and economically by employment in logging, sawmills or pulp mills, a hegemony that became increasingly entrenched in the period after the Second World War. Prudham describes the example of Josephine County in Oregon, where the lumber boom arrived in the 1940s, such that by 1950 the wood products sector accounted for over a quarter of the local workforce, including 90 per cent of all manufacturing employment. Yet, he notes, the 64 lumber mills in the county in 1951 typically employed only eight to ten people. As such, the socio-economic culture of the district was one of petty capitalism:

> Workers outnumbered owners, but by factors of ten, not hundreds. Class distinctions were in significant ways blurred by the fact that owners of small mills worked closely alongside their employees ... Moreover, boundaries between workers and owners were further erased as many of the former tried their hands at entrepreneurship ... returning to wage labor if things did not work out.
>
> (Prudham, 2005: 149)

These socio-economic dynamics led whole communities to identify with the wood products industry and its economic interests. At the same time, however, the territorial communities of the Douglas fir country were also cross-cut by occupational communities reflecting different functions within the industry. Carroll (1995) describes the occupational community of loggers in the Pacific Northwest, self-defined as including anyone 'who builds logging roads or drives log trucks, as well as anyone involved in cutting and moving the logs from the stump to the landing' (p. 67), and set apart from mill workers who were stereotyped by loggers as weak, work-shy and effeminate. In addition to the company loggers, labour for forestry work was also provided by itinerate logging crews and by seasonal tree planters. Thus, a range of parallel rural lifestyles were imagined and performed within the forestry-based communities of the Pacific Northwest, some spatially fixed others mobile, but all defined by engagement in the exploitation of forest land.

Moreover, the forestry industry has regulated and modified the rural landscape of the Douglas fir country. Whilst the industry originally involved the logging of established 'natural' Douglas fir forest, it is estimated that up to 90 per cent of the old-growth forest that had existed in the region in the nineteenth century has now been cleared. Commercial forestry in the Pacific Northwest therefore now depends on industrial reforestation, with 'tree farms' planted for the sole purpose of future harvesting comprising trees that have been selectively bred or genetically engineered to enhance the end product (Prudham, 2005). The physical management of the forest is complemented by a discursive management which by representing the forest as a crop and a commodity has traditionally denied alternative representations, for example, of the forest as a habitat or an ecosystem. In the last three decades, however, these alternative representations have been given new weight and prominence by environmental protests and legislation. Most notably, in June 1990 the United States Fish and Wildlife Service placed the northern spotted owl on its list of endangered species and 'the forest industry's unchallenged supremacy in the Northwest woods ended' (Prudham, 2005: 4). Measures to protect the northern spotted owl in its natural habitat of the Douglas fir forest of the Pacific Northwest involved significant restrictions on the logging of forests owned by US government agencies – closing off access to timber that had been regarded as essential to the economic sustainability of the Pacific Northwest forest industry. With forecasts of tens of thousands of job losses and the implosion of whole communities, the 'spotted owl controversy' became a major political issue in the region, pitting rural economy, society and landscape created by resource capitalism against the rural of biophysical nature discursively constructed by environmentalists.

Expanding the resource frontier

One of the key characteristics of resource capitalism identified by David and Wright (1997) is intensity of search. Resource capitalism relies on seeking out new natural resources for exploitation and, as such, has been a key driver in opening up new areas of peripheral rural land for exploration, settlement and development. From the Klondike to the Congo, history is scattered with failed or short-lived efforts to extract the mineral wealth of remote (and often inhospitable) rural regions. Yet, the potential capital return is such that the discovery of extensive mineral resources in areas such as the arctic and subarctic regions of North America, or Siberia, or the Australian interior, transforms the discursive representation of such spaces from wilderness into places of opportunity.

The development of sustainable mechanisms for exploiting the mineral resources of remote rural regions, however, demands not only substantial capital investment (including public investment in transport infrastructure) and technological innovation, but also the import of labour, housed in company enclosures or new, single-purpose towns that introduced different and distinctive forms of social organization into rural space. The remote mining towns of North America and Australia, for instance, were brought into existence solely to facilitate the exploitation of mineral reserves for capital gain. The town and its inhabitants may have little connection with the surrounding rural environment beyond the mine, which in many cases is largely unsettled and undeveloped. The isolation of such towns, their imported population, and the predominance of manual work and industrial labour relations, often for a single employer, together means that their social structure and culture can appear more urban than rural in character – gaining reputations as islands of trade unionism and left-wing politics, or of vice, including gambling and prostitution, at least historically (see for example, Harvie and Jobes, 2001, on vice in Montana resource towns).

The smooth running of mines and mills as vehicles for capital accumulation, however, demanded the exertion of social and political control over the towns. This need for control favoured the development of instant, company towns, planned settlements created alongside new industrial developments, such as Mackenzie and Tumbler Ridge in northern British Columbia (Halseth and Sullivan, 2002). Mackenzie was established from scratch in 1966, linked to the development of saw and pulp mills and a major hydroelectric project; Tumbler Ridge followed 15 years later to provide labour for new open-cast coal mines. Almost instantly, both towns acquired populations of around 4,000 to 5,000 people. As Halseth and Sullivan (2002) document, the construction of the towns followed principles of progressive planning, providing not only industrial and housing units, but also public services, commercial centres and recreational facilities, each zoned to maintain a separation of industrial and residential land use. Efforts were also put into promoting community interaction, with events sponsored by local government and major employers.

Although Mackenzie and Tumbler Ridge were largely built and shaped by the companies that owned the sawmills and coal mines respectively, in both cases the planned towns were intended to develop economies independent of the core industries and to achieve long-term sustainability. In practice, economic diversification has been limited, especially in Tumbler Ridge, and the towns' fortunes have remained closely tied to their staple industries. In common with other single-industry resource towns, Mackenzie and Tumbler Ridge are highly exposed both to the vagaries of nature (including the finite extent of mineral resources) and to corporate decision-making by companies that have been sold, taken over and amalgamated into global conglomerates. The Quintette Mine at Tumbler Ridge closed in 2000, followed by a 'fire sale' of company-owned housing, whilst Mackenzie has lost a paper mill and two saw mills, and both towns are losing population, but fighting to avoid the fate of single-industry towns elsewhere in Canada such as Cassiar in British Columbia, and Gagnon in Labrador, which were effectively closed down and abandoned once mines closed (Clemenson, 1992; Coates, 2001). The challenge of sustaining resource towns also confronts other peripheral rural regions, including northern Russia, where economic diversification has been attempted with variable success (Tykkyläinen, 2008).

The search for new resource exploitation opportunities continues to push into more and more remote regions, including the extraction of northern Alberta oil sands, the opening of mines in arctic Canada such as Nanisivik mine on Baffin Island and Polaris mine in the Parry Islands, and the prospect of oil drilling in Alaska's Arctic Wildlife Refuge (Bone, 2003; Standlea, 2006). Increasingly the developers of such sites have solved the labour problem by flying in workers on rotational shifts, including indigenous employees air commuting from aboriginal settlements (Bone, 2003). The new model of 'fly in, fly out' development (Markey, 2010), might limit the transformative impact of resource capitalism on remote rural environments, but the targeting of previously 'pristine' wilderness has generated conflicts with environmentalist discourses of rural wilderness, as discussed further in Chapter 9.

THE PRODUCTIVIST COUNTRYSIDE

Introducing productivism

The rise of productivist agriculture reshaped the rural economies, societies and environments of developed market economies during the mid-twentieth century. Productivism refers to a discourse of agricultural organization in which the function of farming was singularly conceived as the production of food and fibre, and which prioritized increasing agricultural production over all other considerations. As such, productivism transformed not only the agricultural industry, but also the whole countryside, requiring the reconfiguration of labour relations, social structures, environmental conditions and landscapes towards support for the singular goal of maximizing agricultural production.

The structural characteristics of productivist agriculture are three-fold (Bowler, 1985; Ilbery and Bowler, 1998). First, it involves the *intensification* of farming, including the mechanization and automation of production processes, the increased use of agri-chemicals to enhance yields and reduce wastage, the application of biotechnology, and the introduction of 'factory farming' methods of livestock rearing. Second, productivist agriculture involves *concentration* within the agricultural sector, with the amalgamation of fields and farms, the expansion of agri-business and corporate farming, increasing reliance on producing under contract to food processors and retailers, and the streamlining of the food commodity chain through corporate mergers and alliances. Third, it also involves *specialization* within farming, including the expansion of monoculture as individual farms specialize in particular products and the concurrent emergence of patterns of regional specialization, as well as labour specialization and the increased use of contract labour and agricultural services. Each of these structural dimensions of productivism has had consequences both for the practice of farming and for the wider social, economic and environment dynamics of rural areas, as Table 3.1 summarizes.

The development of productivist agriculture in the mid-twentieth century was stimulated and conditioned by factors both within and beyond the farming sector. Ilbery and Bowler (1998) outline three competing theorizations that have been advanced to explain the rise of productivism, each of which position agriculture within a broader socio-economic framework. Thus, the first model of *commercialization* emphasizes the significance of economic factors in driving the modernization of farm production processes and the integration of farms into capitalist market relations. Strongly influenced by neoclassical economics, the commercialization model holds that family farms are transformed into capitalist enterprises by the introduction of supply–demand relations in a market economy. Productivist techniques and principles are hence adopted to help farms meet market demand and, moreover, to maintain competitive advantage within the marketplace. In particular, the commercialization thesis stresses the importance of technological innovation in supporting agricultural development, arguing that technology is 'a cause (not an effect) of agricultural modernization and economic development' (Ilbery and Bowler, 1998: 58). However, it is also argued that commercialization produces a 'dual farming economy', with divergence between modern capitalistic farms and traditional family farms. The mix of these farm types varies spatially, with distance from major urban markets, regional farm size structures, and the distribution of natural resources employed to explain the differential geography of agricultural modernization.

The second model, *commoditization*, adopts a political-economy approach and emphasizes the significance of social structures over economic structures and of farm outputs over farm inputs. It proposes that farm households in modern, capitalist society have become increasing dependent on goods obtained through the market, and thus are themselves required to sell through the marketplace in order to earn income to purchase farm inputs (Ilbery and Bowler, 1998). Farms hence get drawn into the capitalist accumulation process, which in turn restructures the nature of farm labour. Modern capitalist farms develop in which there is a separation between the capitalist landowning class that owns and manages the farm, and the working class that provides labour in return for wages. Traditional

Table 3.1 Structural dimensions, process responses and consequences of the industrialization of agriculture (after Ilbery and Bowler, 1998)

Structural dimension	Primary process responses	Secondary consequences
Intensification	Purchased inputs replace labour and substitute for land, increasing dependence on agro-inputs industries Mechanization and automation of production processes Application of developments in biotechnology	Development of supply (requisites) cooperatives Rising agricultural indebtedness Increasing energy intensity and dependence on fossil fuels Overproduction for the domestic market Destruction of environment and agro-ecosystems
Concentration	Fewer but larger farming units Production of most crops and livestock concentrated on fewer farms, regions and countries Sale of farm produce to food processing industries – increasing dependence on contract farming	Development of marketing cooperatives New social relations in rural communities Inability of young to enter farming Polarization of the farm size structure Corporate ownership of land Increasing inequalities in farm incomes between farm sizes, types and locations State agricultural policies favouring large farms and certain regions
Specialization	Labour specialization, including the management function Fewer farm products from each farm, region and country	Food consumed outside region where it was produced Increased risk of system failure Change composition of the workforce Structural rigidity in farm production

family farms, dependent on family labour, it is argued, cannot compete effectively with capitalist wage-labour farms, and should, according to commoditization theory, be gradually squeezed out of existence. However, although the commoditization thesis enjoyed some popularity in the 1980s, the empirical evidence for the thesis is weak. In particular, critics have argued that the continuing presence of family farms in advanced capitalist societies disproves the assumptions of the model (see Ilbery and Bowler, 1998).

The third model, *industrialization*, combines elements from both the commercialization and the commoditization approaches but also adopts a wider perspective in examining the larger food-supply system and its long-term development. In particular, the industrialization approach focuses on the transformation of agriculture by industrial capital over time, and the conditioning of this transformation by responses to bio-physical and natural processes. The embeddedness of agriculture in nature, it is argued, means that changes in agricultural production are not uniform and hence that the development of capitalist agriculture has proceeded in a series of steps (Goodman *et al.*, 1987; Ilbery and Bowler, 1998). Thus, for example, the replacement of animal power by machines was an early step, followed later by the replacement of human labour by machines. Goodman and Redclift (1991) further argued that this incremental evolution of agriculture involves both processes of appropriation and processes of substitution. Appropriation here refers to the removal from the farm production process of elements that could be naturally reproduced on their farm and their replacement by commercially purchased inputs. The replacement of animals by farm machinery and of manure by synthetic fertilizers are examples of this. Substitution, in this context, refers to the replacement of cultivated food and fibre by industrially produced substitutes, especially in the clothing and food processing industries, thus changing demand for agricultural products. Examples of substitution include the development of artificial sweeteners to replace sugar in foods, and the use of nylon in place of cotton.

The industrialization thesis therefore acknowledges that changes in agriculture are shaped not only by farm-level decision-making, but also by wider social, economic, political and technological dynamics impacting on the food-supply system. Goodman and Redclift (1989) describe agricultural production as positioned within historically specific *food regimes* which have fixed relations between food production and consumption on a global scale in a sequence extending back to the 1870s. Productivism is identified with the second food regime, prevalent from the 1920s to the 1980s, which involved the transnational restructuring of agriculture in order to supply increasingly global mass markets (Le Heron, 1993).

As such, the industrialization approach affords the opportunity to consider how wider social and political events and relations have shaped the evolution of food regimes, and, within them, the development of capitalist agriculture. In particular, it allows us to recognize the role of the state in supporting and sponsoring productivism, through the funding of research and development programmes and agricultural training; the award of grants, subsidies and loans to farmers to enable investment in farm modernization; the regulation of markets by price support mechanisms; and the protection of domestic markets by tariffs and import restrictions. These measures were enshrined in state policies, most notably the Common Agricultural Policy of the European Community, and afforded agriculture a degree of state favour that was arguably unmatched for any other industry – a situation that has been labelled 'agricultural exceptionalism'.

The ease with which agriculture was able to command substantial state resources in support of the productivist project reflected the prevalence of concerns in the mid-twentieth century about the need to feed a rapidly growing and increasingly urban population – concerns that were compounded in Europe by the disruptive impact of the Second World War on agricultural systems. Agricultural capitalists and state agents were hence able to articulate a discourse that prioritized maximizing agricultural production as the solution to the food crisis. In this discourse, rural areas came to be positioned primarily as sources of food, and the constituent elements of rural life, from livestock to farmworkers to hedgerows, were represented as components in the agricultural mode of production to be modified, refined or replaced as appropriate.

The implementation of productivist policies brought substantial material benefits for agricultural capitalists, however, in several respects state intervention served to moderate and constrain the capitalist accumulation process – for example, through the use of artificial market mechanisms to control prices for agricultural products, the establishment of state-owned purchasing and marketing monopolies for key commodities, and the regulation of agricultural wages. Indeed, the survival of the family farm in Europe may be credited to the protection provided to small farmers from full exposure to free market conditions by the Common Agricultural Policy, whilst the rolling back of state intervention in agriculture in Australia and New Zealand has resulted in the disappearance of many smaller, family-run farms (see Chapter 8).

Furthermore, whilst the commercialization, commoditization and industrialization models all position productivism as part of the development of capitalist agriculture in advanced market economies, it might be noted that productivist principles were also adopted in agricultural production in state socialist countries, and that parallels could be found in the 'Green Revolution' experienced in developing nations such as India and Mexico (see Box 3.2).

Box 3.2 THE GREEN REVOLUTION

First world productivism, with its drive to increase agricultural production, had a parallel in the Green Revolution experienced by many parts of the global south during the 1960s and 1970s. In common with European productivism, the Green Revolution was driven by the urgent need to secure food supplies for a booming population. Yet, not only did the scale of the projected population increase far outstrip the post-war 'baby boom' of Europe, but the Green Revolution also needed to address problems of endemic extreme poverty in rural areas, basic agricultural technologies and techniques, vulnerability to disease and extreme weather, and the adaptation of post-colonial economies. Many governments also enthusiastically backed the Green Revolution as a political strategy to support post-colonial goals of national self-sufficiency and to limit demands for more radical land redistribution and economic reforms (Atkins and Bowler, 2001).

The Green Revolution involved the enhancement of agricultural production in developing countries through mechanization, irrigation, improvements in land management, the introduction of pesticides and other agri-chemicals, and, most notably, the development of new hybrid varieties of rice and wheat that could resist certain diseases, such as the rust virus in cereals, and produce higher yields (Atkins and Bowler, 2001; Southgate *et al.*, 2007). Collectively, these innovations resulted in dramatic improvements in agricultural output, including a 27 per cent increase in per capita food production in Asia between the mid-1960s and mid-1990s (Jewitt and Baker, 2007), and have helped to avoid the threat of severe famine in regions such as the Indian sub-continent. The Green Revolution also had a transformative impact on broader aspects of rural life. Social relations were reconfigured as permanent agricultural employment expanded, creating a new rural proletariat that challenged traditional class and caste structures in countries such as India. In a study of three Indian villages in 2001, following-up research in 1972, Jewitt and Baker record improvements to infrastructure, ownership of consumer goods, health and literacy as well as to food security, that are attributed by villagers to the Green Revolution:

> open wells had largely been replaced by domestic hand pumps, brick houses has replaced mud-built ones and cars and tractors were numerous. With respect to the villages themselves ... standards of dress were much higher, the eye infections and septic wounds so frequently evident on limbs in 1972 were virtually absent and most people *seemed* a great deal healthier and better fed.
>
> (Jewitt and Baker, 2007: 78)

However, as with European productivism, the Green Revolution has been accused of prioritizing agricultural production over wider social and environmental interests. Rural landscapes have been transformed as indigenous crops have been replaced by high-yield hybrid varieties, with biodiversity reduced, and serious concerns have been raised about the environmental impact of agricultural intensification (Jewitt and Baker, 2007; Southgate *et al.*, 2007). The social impact is also complex, with criticism that the Green Revolution increased social inequalities within agrarian communities (Baker and Jewitt, 2007). Furthermore, the substitution of

continued

indigenous crops by commercially produced hybrid seeds has arguably left farmers dependent on transnational biotechnology corporations and exposed them to pressure to adopt controversial genetically modified varieties (Richards, 2004). Thus, whilst Jewitt and Baker (2007) identify grassroots acceptance of the benefits of the Green Revolution, a strong post-development critique has been advanced by critics, including the Indian activist writer Vandana Shiva, who has argued that the 'large scale experiment of the Green Revolution has not only pushed nature to the verge of ecological breakdown, but also seems to have pushed society to the verge of social breakdown' (Shiva, 1991: 172).

Further reading: Jewitt and Baker (2007).

The productivist paradox

The capacity of productivist agriculture to transform the rural areas of developed market economies resulted from the combination of its discursive and material effects. Discursively, productivism reiterated and reinforced the representation that the primary function of rural areas is to produce food, fibre and other natural resources. This foundational principle underpinned the development of productivist policies and justified the exclusion or dismissal of alternative, competing representations that questioned the appropriateness of productivist techniques and objectives, or highlighted other non-productive functions of rural land.

The hegemony of the productivist discourse in government policy was replicated in the lay discourses of many rural residents, especially within the farming community itself. As Burton (2004) observes, productivist rationalities became, and remain, intrinsic to farmers' self-conception of their identity, which they construct primarily in terms of the production of food and fibre. Burton's interviews with farmers in southern England demonstrate that perceptions of what it means to be a 'good farmer' are connected, at least in part, with the ability to increase production and maximize crop yield. Thus, the 'good farmer' was described by one farmer as 'the chap who can up his output by a ton an acre or whatever – and continue to do so' (Burton, 2004: 202), and by another as someone who 'tries to get three heads of corn where there used to be two, or three blades of grass where there used to be two' (*Ibid*: 203). A third farmer suggested that a good farmer would be:

> always looking to produce more per acre than already produced. It's the aim of everyone ... at least it should be if you're a proper farmer.
>
> (quoted by Burton, 2004: 202–3)

Adherence to these values was enforced and reproduced through the social interaction of farmers, talking and boasting about the size of crop yield in the pub, and through the visual surveillance of farmers observing each other's fields over hedges and fences. Surveillance and peer pressure could act to guard against the more excessive or careless application of new farming technologies or agri-chemicals, reflecting the farmers' penchant for landscapes that are 'neat, clean and ordered'. Yet, it is also clear from Burton's study that farmers' perceptions of neatness and orderliness are framed within productivist rationality, such that, for example, the spraying of fields with chemical herbicides is regarded as good husbandry. As Burton notes, many of the practices that are given high symbolic value by farmers also produce economic benefits, although they might not in themselves be an accurate indication of profitability.

The dominance of productivist discourses at all levels of rural society normalized the drive for increased agricultural production and legitimized changes to farming practice that would have dramatic material consequences for the economy, society and environment of rural areas. The material effects of productivism are perhaps most readily evident in its environmental impact. Large parts of the rural landscape were transformed by the amalgamation of farm units, the removal of hedgerows and fences create larger fields, and the eradication of ponds, marshes, copses and other fragments of unproductive land. Over 200,000 kilometres of hedgerow were removed in England and Wales, for instance, in the five decades after the 1947 Agriculture Act ushered in the productivist era, reducing the total extent of hedgerows by almost a quarter (Green, 1996; Woods, 2005a). The appearance of the landscape has also been changed by the conversion of pasture, orchards and meadows to arable land to chase more profitable returns. An estimated 97 per cent of wild flower meadows in Britain have been lost since the 1960s through a combination of conversion to arable land, the application of herbicides and poor land management (Green, 1996; Woods, 2005a).

The destruction of habitats and the increased use of chemical pesticides, herbicides and insecticides additionally had the effect of reducing the numbers of birds and other wildlife resident in the agricultural landscape. The population of 12 common farmland bird species in England, for example, fell by 58 per cent between 1978 and 1998 (Woods, 2005a). Furthermore, productivist agriculture has been associated with problems of soil erosion and flooding, the depletion of aquifers and the pollution and eutrophication of watercourses (Green, 1996; Harvey, 1998; Woods, 2005a). Yet, for most of the post-war period, the strength of the productivist discourse was such that the environmental consequences of productivist agriculture were obfuscated, dismissed or constructed as 'non-issues'. Whilst high-profile revelations such as Rachel Carson's (1962) exposé of the lethal legacy of the pesticide compound DDT (eventually) forced some changes, other effects were either tolerated as the acceptable cost of increased agricultural production (for example, hedgerow removal) or internalized within the farm community and exempted from normal environmental regulation (for example, farm effluent pollution) (Lowe *et al.*, 1997).

The social and economic effects of productivist agriculture for rural areas have been no less significant. In particular, labour specialization and the mechanization of farm production processes substantially reduced the agricultural workforce in developed market economies. In Britain, the number of hired farmworkers fell from over 800,000 in the 1940s to under 300,000 in the 1990s (Clark, 1991; Woods, 2005a). In France, the agricultural workforce, including farmers, decreased from over five million in 1954, to around two million in 1975 (Woods, 2005a). Many former farmworkers migrated away from rural communities to find employment, contributing to a dominant trend of rural depopulation in the early post-war era. Ironically, the deserted cottages and farm houses that were left behind were later often bought and gentrified by in-migrants from towns and cities as the migration tide reversed (see Chapter 4).

The former farmworkers that remained in rural areas moved into employment in new manufacturing and service industries (often attracted as part of rural development projects aimed at replacing lost agricultural employment), which in many areas rapidly surpassed farming in terms of numbers employed. At the same time, the significance of agriculture to local rural economies was further diminished by the trend of farms producing directly under contract to food processors and retailers, by-passing local firms.

The consequence of these changes is that ties between farmers and rural communities have become weakened and the overshadowing presence of agriculture in rural life has been eroded. Farmers no longer dominate positions of community leadership as they once did in the rural areas of countries such as England (Woods, 2005b), and key points in the farming calendar such as lambing and harvest are no longer events that engage the whole rural community. The closure or rationalization of rural services such as village schools, post offices and pubs has limited the opportunities for interaction between the farm and non-farm population (Errington, 1997). Recent studies in England have suggested that only around a third of farmers consider themselves to be 'very actively involved' in the local community (Lobley *et al.*, 2005), and that around one in eight rural residents do not have any social contact with farmers (Milbourne *et al.*, 2001).

As the farm population has become more detached from the wider rural community, tensions and conflicts between farmers and non-farmers are perceived to have increased, especially with the influx of ex-urban incomers with little background knowledge or experience of agriculture. Courtney *et al.* (2007), for example, report that 10 per cent of the farmers that they surveyed in case study localities in England had received complaints from non-farming residents, notably in relation to slurry on roads and impediments to public access to footpaths. Yet, studies also indicate that differences in perception between the farm and non-farm populations are more significant than actual incidents of conflict. Research by Smithers *et al.* (2005) in southern Ontario, Canada, for instance, found that both farmers and non-farmers acknowledged the significance of agriculture to the economy and character of the locality, but that opinion diverged when asked about the major challenges facing farming. Most non-farmers identified environmental concerns as the major challenge for agriculture, followed by competition from factory farms. The farmers surveyed, however, most frequently identified low commodity prices as the major challenge, with environmental concerns given less emphasis. Smithers *et al.* (2005) report that over three-quarters of the farmers questioned believed that local townspeople were largely unaware of what went on in local agriculture and the problems faced by farmers and that their opinions were shaped by more generalized impressions of farming. As one farmer commented, 'some still think that you have old Macdonald's farm here ... unless they've come from a farm they don't understand farming ... they think they do, but they don't' (Smithers *et al.*, 2005: 288).

This is the productivist paradox. Productivist agriculture is founded on the idea that the primary purpose of rural land is the production of food, fibre and other farm outputs, such that the drive to increase agricultural production could be legitimately prioritized above all other concerns. Yet, the practice of productivist agriculture had the effect of weakening the actual influence of agriculture in rural communities, thus creating the conditions in which alternative discourses could be articulated to challenge the exceptional status afforded to agriculture. As such, productivist agriculture can be claimed to have created the circumstances of its own downfall.

The post-productivist transition?

Over the four decades that followed the end of the Second World War, productivist policies were outstandingly successful in achieving their aim of increasing agricultural production in the global north. Total production of cereals in Europe, North America, Australia and New Zealand, for example, increased by 75 per cent between 1961 and 1981; whilst production of meat in these regions increased by 71 per cent over the same period. Indeed, the increase in productivity was so successful that by the end of the 1970s, agriculture in the global north was overproducing against market demand in a number of commodities. The capacity of the agricultural sector to naturally adjust to the problem of over-production, however, was limited by the artificial market control mechanisms that had been introduced as part of the productivist revolution. In particular, price support mechanisms which involved state intervention to buy surplus goods once commodity prices hit a specified low price meant that there was no incentive for farmers to avoid over-production; whilst protectionist trade policies limited opportunities for selling surplus product on world markets. Accordingly, large quantities of surplus commodities were simply stored in expanding stockpiles. In 1982, the European Community was storing nearly seven million tonnes of surplus wheat; by 1984, it has also accumulated stockpiles of over one million tonnes of butter and around 500,000 tonnes of beef carcasses (Winter, 1996).

Whilst state investment in productivist agriculture had been backed by political consensus in the post-war era when concerns about food supply predominated; the continuing expenditure of large volumes of state finance to support agricultural over-production became increasingly controversial, especially in the climate of fiscal restraint in state expenditure that pervaded in the early 1980s. Additionally, growing public concerns about the environmental impact of intensive farming, animal welfare and food quality created a groundswell of pressure for reforms in agricultural policy and practice. In Australia and New Zealand, radical reforms dismantling state support for agriculture were implemented, as discussed further in Chapter 8. In the European Union and the United States, in contrast, the farming lobbies were strong enough to frustrate efforts at radical reform, but tentative steps away from productivism were made through a series of more modest and gradual measures that became known as the 'post-productivist transition'. These included encouragement for farm diversification, creating a second non-agricultural income stream for farm households, for example through tourism or recreation; payments to farmers for environmental improvements, including planting woodland and leaving fields fallow; support for initiatives aimed at producing high-quality locally branded food; and assistance with conversion to organic farming.

However, whilst reference to the 'post-productivist transition' is useful in indicating a change in direction of government policy, its wider application in rural studies and its transmogrification into a concept of 'post-productivism' have been controversial. Through the 1980s and 1990s the term increasingly became used by rural geographers to suggest the emergence of a new era of agricultural production. Yet, Robinson (2004) argues that the concept of 'post-productivism' was adopted too quickly and too uncritically, and a number of commentaries have critiqued the approach, presenting two key charges. First, the concept of 'post-productivism' was theoretically weak and poorly defined. Different authors endowed it with different characteristics, such that there was no coherent framework for identifying a 'post-productivist countryside' (Evans *et al.*, 2002; Ilbery and Bowler, 1998; Mather *et al.*, 2006; Wilson, 2001). Second, although the loose and general usage of the term suggested a significant transformation of agricultural practice away from productivist ideals, empirical studies presented limited evidence to support this assertion (Argent, 2002; Evans *et al.*, 2002; Walford, 2003). As such,

Evans *et al.* (2002) have argued that 'more progress in agricultural (and rural) geography could be achieved by abandoning post-productivism' (p. 326). Rural geographers have hence started to search for alternative ways of conceptualizing change in the rural economy, with attention particularly focusing on the increasingly multi-functional nature of the contemporary countryside.

TOWARDS A MULTIFUNCTIONAL COUNTRYSIDE?

Multifunctional agriculture

The concept of 'multifunctionality' originates in attempts by rural geographers and rural sociologists to move beyond the apparent deadlock of the dichotomy of 'productivist' and 'post-productivist' agriculture that had been reached by the late 1990s. Critics such as Marsden (1999, 2003) and Wilson (2001, 2007) responded to the conceptual and empirical difficulties that had been identified with 'post-productivism' by challenging the implication that post-productivism was something that happened *after* productivism in a linear progression. Instead, the notion of the 'multifunctional agricultural regime' was introduced, allowing for 'the multidimensional coexistence of productivist and post-productivist action and thought' (Wilson, 2001: 95), and which, it was argued, provided 'a more accurate depiction of the multi-layered nature of rural and agricultural change' (*Ibid*).

However, as the concept of multifunctionality has been adopted more widely within rural geography and rural sociology, it has come to mean more than simply a differentiated agricultural economy. In particular, it has come to refer to the multiple outcomes of agriculture, which include not only the production of food and other resources, but also social and environmental benefits. As Potter and Burney summarize:

> agriculture is multifunctional, producing not only food but also sustaining rural landscapes, protecting biodiversity, generating employment and contributing to the viability of rural areas.
>
> (Potter and Burney, 2002: 35)

Framed in this way, multifunctionality serves not only as a mechanism for describing and analysing the mixture of productivist and post-productivist practices evident in a territory, but also is a model for understanding the dynamics of agricultural systems. As such it has assumed a normative meaning, championed by family farm and environmental campaigners, as well by some advocates of agricultural liberalization, as an objective for agricultural policy (Potter, 2004; Potter and Tilzey, 2005).

There are consequently at least three linked yet different inflections of multifunctionality currently employed in rural geography research. In the first, multifunctionality is viewed as a policy outcome, with the focus of investigation directed towards analysis of agricultural policies and the extent to which they recognize and support the multiple functions of agriculture (Bjørkhaug and Richards, 2008; Potter, 2004; Potter and Burney, 2002; Potter and Tilzey, 2005). In the second approach, multifunctionality is presented as 'a complex transition comprised of weak, moderate and strong multifunctionality pathways' (Wilson, 2008: 367) that can be applied to explain trajectories of change on individual farms, moving between poles of productivist and non-productivist action and thought (see also Wilson, 2007). In the third approach, multifunctionality is seen as a characteristic of national or regional agricultural regimes, focusing on the combination of farming practices across a territory (Holmes, 2006) or the holistic practices and functions of a regional industry (Hollander, 2004). Arguably, the last approach can be extended beyond agriculture to other natural resource industries, such as forestry, as well as to the holistic operation of regional rural economies in which production- and consumption-oriented activities are entwined and inter-dependent.

Whichever of these approaches is adopted, multifunctionality necessarily implies a rethinking of the meaning and purpose of rural space, moving away from the productivist logic that framed the rural as an industrial site for resource production and exploitation. Yet, multifunctionality is not in this respect as radical as its proponents might imply. Multifunctional agricultural regimes are still centred on the exploitation of the land through farming, and are still located within a capitalist paradigm in requiring the commoditization of agricultural goods and benefits. Morever, they also acknowledge the continuing significance of productivist agricultural practices, permitting, and arguably prioritizing, industrialized farming for mass production, where this can be supported through the market. Where multifunctionality differs from previous approaches is on the question of what happens to farms that cannot be viably sustained through the free market for agricultural produce. Multifunctionality recognizes that such farms have a value to the countryside over and above their production of goods for the mass market, and seeks to enable these wider functions to be valorized in order to achieve economic sustainability. This might mean converting to the production of higher-value agricultural goods such as organic food, or exploitation of the amenity value of farmland through tourism and recreational activities, or the commodification of the environmental benefits of farming through the payment of rewards for good stewardship (Plate 3.2). In this way, the transition to multifunctional agricultural regimes can be positioned as part of the wider shift in the global economy from Fordist regimes to post-Fordist regimes, from mass production and standardization to

Plate 3.2 Multifunctionality on a farm in the Netherlands: a nature reserve and public paths combined with sunflower cultivation (Photo: author)

specialization and the development of niche markets (Potter and Tilzey, 2005).

Super-productivism and the global agri-food complex

As the concept of multifunctionality has debunked the misinterpretation of the post-productivist transition as a move away from productivism in general, light has been thrown not only on the strategies of farmers attempting to diversify into non-productivist activities, but also on the continuing intensification of productivist practices within global agriculture. Indeed, it can be argued that the search for non-productivist alternatives embodied in the concept of multifunctionality stems not from the failure of productivism, but from its success in streamlining the agricultural production process. As has been the case in other industries, the efficient high-volume production of food and fibre through farming has become spatially concentrated in sites where costs can be minimized and production maximized to serve mass markets on a global scale.

The processes involved in this advanced refinement of agricultural production, described by Halfacree (1999, 2006) as 'super-productivism', builds on principles and techniques that have been central to capitalist agriculture since at least the resource-boom of late nineteenth-century California. First, there is a continuing effort to improve the environment for agricultural production and reduce the threat from natural externalities (poor weather, drought, predators, and so on). Long-established techniques such as irrigation continue to be rolled out, with the area of irrigated farmland in the global south projected to increase by 20 per cent between 1999 and 2030 (Bruinsma, 2003). 'Factory farming' of livestock has also intensified, with 74 per cent of the world's poultry and 50 per cent of pig-meat produced through concentrated indoor rearing (Weis, 2007), generating dystopian accounts of industrial landscapes and assembly-line processing as in Cockburn's description of pig-farming in North Carolina:

> Its reeking lagoons surround darkened warehouses of animals trapped in metal crates barely larger than their bodies, tails chopped off, pumped with corn, soy beans and chemicals until, in six months, they weigh about 240 pounds, at which point they are shipped off to abattoirs to be killed.
>
> (Cockburn, 1996: 39)

Public concerns about the animal welfare and environmental consequences of industrialized farming have resulted in increased regulation in developed nations, including 'proposition 2' passed by voters in California in November 2008 restricting the confinement of farm animals. Yet, the intensive indoor rearing of livestock is expanding rapidly in the global south, supporting dramatic increases in meat consumption in countries such as China and India.

Similarly, hydroponic farming techniques have facilitated the substantial expansion of under-cover climate-controlled crop cultivation by utilizing new technologies that feed plants with mineral nutrient solutions, enabling them to grow without soil. The landscape of Almería in southern Spain has been transformed by 350 km^2 of plastic greenhouses in which tomatoes, courgettes and other crops are cultivated using such technologies (Tremlett, 2005). In Britain, the same hydroponic techniques will be employed in a new 'super-farm', Thanet Earth, comprising seven extensive greenhouses covering 92 hectares with 1.3 million plants growing suspended in rows from the ceiling. This single enterprise is forecast to increase Britain's total salad crop by 15 per cent (Addley, 2008). New technologies are also employed in precision farming, in which data from automatic sensors and satellite monitoring is combined with GIS models and global positioning systems (GPS) to micro-manage crops grown across large farms, for example, by concentrating fertilizer use.

Second, research and development in biotechnology has continued to 'improve' livestock and crops. In particular, since the early 1990s this has controversially meant using genetic engineering to modify the properties of crops and farm animals. In many cases, genetically modified organisms (GMOs) have been developed that can increase resistance to insects or viruses, or tolerance to herbicides, but GMOs are also being developed to increase yields and to improve the health qualities of food, to change the colour and shape of produce, and to meet other specific market demands. GM crops and livestock clearly offer benefits for agrarian capitalism, but advocates of GM agriculture also promote it as a means of developing more resilient crops for use in developing countries, and emphasize the projected financial gain for developing economies (Anderson and Valenzuela, 2008; Moschini, 2008).

Yet, there are also widespread fears about the untested environmental impact of GM agriculture, and even about the safety of GMOs for human consumption. Public opposition has restricted the use of GMOs in agriculture in many developed countries, particularly in Europe, and farmers have actively resisted GM crop adoption in parts of India (see Hall, 2008; McAfee, 2008; Pechlaner and Otero, 2008; Scoones, 2008). As such, although the area of land under GM cultivation has climbed steeply, increasing by over 375 per cent between 1997 and 2001 (Bruinsma, 2003; Millstone and Lang, 2003), and by a further 117 per cent to 2007, the geography of GM agriculture is steadfastly polarized (Table 3.2). Only 13 countries had permitted the commercial production of GM crops by 2001, with 91 per cent of GM crops grown in the United States and Argentina. With the expansion of GM agriculture continuing to face

Table 3.2 Distribution of commercial GM crop cultivation, 1996–2007

	Area under commercial GM crops (million hectares)			
	1996	**1999**	**2001**	**2007**
United States	1.5	28.7	35.7	57.7
Argentina	0.1	6.7	11.8	19.1
Brazil	0	0	0	15.0
Canada	0.1	4.0	3.2	7.0
India	0	0	0	6.2
China	1.1	0.3	1.5	3.8
Paraguay	0	0	0	2.6
South Africa	0	0.1	0.2	1.8
Uruguay	0	<0.1	<0.1	0.5
Philippines	0	0	0	0.3
Australia	<0.1	0.1	0.2	0.1
Other developed countries[1]	0	<0.1	<0.1	<0.1
Other developing countries[2]	<0.1	<0.1	<0.1	<0.1
Global total	1.7	39.9	52.6	114.3

Sources: Bruinsma (2003), ISAAA (2007), Millstone and Lang (2003)

1 Includes France, Germany, Portugal, Romania and Spain in 2001, plus Czech Republic, Slovakia and Poland in 2007

2 Mexico in 2001, plus Chile, Colombia and Honduras in 2007

obstacles in Europe, Canada and Mexico, biotechnology corporations are increasingly targeting developing nations in Latin America, South East Asia and Africa, where the potential for organized public opposition is more limited, with resulting large increases in GM cultivation in countries such as Brazil, India and Paraguay (Paul and Steinbrecher, 2003).

Third, super-productivism is associated with further corporate concentration and shifting corporate spatial strategies. The search for efficiency gains has produced corporate mergers and strategic alliances that create both horizontal integration – linking equivalent operations in different countries – and vertical integration – connecting the different stages of the food chain. Hendrickson and Heffernan (2002), for example, describe the complex networks comprising three of the largest 'food chain clusters', stretching from biotechnology firms to supermarkets. As state intervention in agriculture has been cut-back, it is these global corporate networks that increasingly exert influence over the agri-food system 'from seed to shelf', to quote the corporate slogan of ConAgra (see also Kneen, 2002).

The spatial strategies of agri-food corporations have also been reconfigured to the global scale. This includes the search for new markets, with lobbying for global free trade (discussed further in Chapter 8), and the promotion of GMOs in countries such as India as a vehicle for securing farmer dependency on proprietary hybrid seed (Kneen, 2002). Production has also been re-targeted on a global scale, as new technologies have allowed agricultural industries to become 'footloose', relocating to regions where costs are lowest or regulations most permissive. Inevitably, the targets have been economically depressed rural regions in both the global north and the global south, where local political authorities are prepared to subordinate environmental concerns to opportunities for job creation and income generation. Quarlman (2001) and Ramp and Koc (2001), for instance, describe the rapid expansion of industrial hog farming in Alberta and Manitoba in western Canada, and the associated development of a major pork-processing plant in Lethbridge, Alberta, primarily to serve Asian markets.

Super-productivism has therefore extended the logic of trends that were evident in post-war productivism and in earlier expressions of resource capitalism, but in doing so it has shed the moral dimension that was still present in productivism – that agriculture involved the improvement of the land, and that agrarian capitalism offered the best way of supporting rural livelihoods and sustaining rural cultures. In contrast, the super-productivist vision of the countryside positions rural land 'solely as a productive resource linked to profit maximization' (Halfacree, 2006: 57). Indeed, it might be questioned whether super-productivist industrial agriculture still has an intrinsically rural character. With the use of precision farming technologies eliminating dependency on local rural environmental conditions, genetic engineering, standardization and landscapes of enclosed livestock units and outsized greenhouses, the continuing rural location of industrial agriculture appears to be down largely to the demand for space and lingering convention.

Back to the land?

The frame of multifunctionality accommodates many different pathways that farmers are now following. Thus, just as super-productivist agriculture has further removed processes of agricultural production from their embeddedness in rural environments, cultures and societies – and further distanced food consumption from food production (see Gouveia and Juska, 2002) – many farms seeking to turn away from the productivist model have moved in the opposite direction. A prominent feature of multifunctional agricultural regimes is the re-assertion of forms of farming that present agriculture not just as a means of capital accumulation, but also as a component within a wider rural system. Such alternative models of agriculture involve a very different representation of the rural than that articulated in super-productivism, with rural space not acting solely as a site for accumulation through resource exploitation, but understood as a space constituted by interlocking social, cultural, environmental and economic elements in which the responsible exploitation of natural resources is a key mediating practice.

Organic farming, which involves a rejection of many contemporary biotechnologies, notably agri-chemicals, and emphasizes an ecological view of the farm system, is one of the most high-profile alternative models for agriculture. Organic food has gained popularity with consumers concerned about food safety and animal welfare, with worldwide sales of certified organic produce increasing from US$10 billion in 1997 to US$17.5 billion in 2000 and US$33 billion in 2005 (IFOAM, 2007; Millstone and Lang, 2003). The growing market demand, the premium prices paid by consumers for organic goods, and the availability of state subsidies for organic conversion, have encouraged farmers to move into organic agriculture. The total area of certified organic farmland in the European Union, for example, increased by 215 per cent between 1995 and 2000 (Millstone and Lang, 2003). Furthermore, whilst certified organic sales are largely concentrated in Europe and North America, and the majority of certified organic land is located in the global north, it is estimated that around 80 per cent of farms in the global south are in effect organic and would not need to change their farming methods to achieve certification (*Ibid*).

However, the expansion of organic agriculture has produced strains within the organic movement. Some commentators have argued that organic farming is a social movement as much as a production regime (Reed, 2008; Tovey, 1997), embodying values of environmental sustainability, working with nature, and linking food production and consumption, and emphasizing small-scale production within localized economies (Guthman, 2002; Morgan and Murdoch, 2000). Several early pioneers were additionally associated with radical social politics and anti-modernity ideologies (Matless, 1998). As such, the vision of the countryside presented by the organic movement is far removed from that articulated within productivism, with very different material outcomes for the organization of rural society and the exploitation of the rural environment. Yet, the principles of the organic movement militated against mass production and restricted its ability to respond to growing consumer demand. Market demand has hence been increasingly met by the engagement of mainstream agri-business in organic agriculture, described as the 'conventionalization' of the sector (Buck *et al.*, 1997; Guthman, 2004).

Guthman (2004) fittingly details the conventionalization of organic agriculture in California, where we started this chapter. As she records, California is the epicentre of American organic agriculture, with its early development stimulated by the influence of urban counter-culture, and with total organic sales in the state in 2002 reaching US$263 million. Yet, California has also witnessed significant agri-business penetration into the organic sector, with companies introducing industrial farming techniques where possible, adopting aggressive marketing strategies, and securing a global presence through the take-over of smaller organic operations. Through these actions, Guthman argues, agri-business has appropriated the organic brand, setting conditions that 'undermine the ability of even the most committed producers to practice a purely alternative form of organic farming' (pp. 301–2).

Guthman acknowledges that the presence of agri-business in Californian organic farming in part reflects the exceptional history of Californian agriculture, as described at the start of this chapter, but as with resource capitalism, the extreme case of California illustrates wider truths. Conventionalization is evident in organic farming in other geographical contexts, driven not only by agri-business but also by supermarkets. The significance of supermarkets as the major vehicle for organic food sales in Britain, for example, has led to a squeeze on farm-gate prices similar to that experienced in conventional agriculture (Smith and Marsden, 2004). As organic farming is increasingly seen as an alternative accumulation strategy for agrarian capitalism – rather than as an alternative framework for working the rural environment – the capacity to sustain premium prices is critical to its continued expansion. This may prove challenging in a global economic downturn. Sales of organic food in Britain fell sharply in the first quarter of 2008 (Jowitt, 2008), and the area of organically farmed land in Britain has decreased from a peak in 2004 (Defra, 2009).

CONCLUSION

This chapter has examined the shifting regimes through which the rural has been exploited as a source of food, fuel and other natural resources, within the dominant framework of the capitalist mode of production. From the pioneer resource capitalists of late nineteenth-century California to the isolated mines of arctic Canada, from colonial plantations to modern organic farms, the physical, biological and human components of the rural economy have been organized, modified and exploited for profit. These activities have been underpinned by a prevailing discourse of the rural as a space of production, used to justify the prioritization of the interests of agricultural and resource capitalism over all other social and environmental concerns, but they also have had substantial material impacts – transforming landscapes, changing habitats and ecological systems, creating new labour and social relations, reconfiguring settlement patterns and constructing transport networks, and founding new institutions. Features that we now consider to be icons of rural life, from the family farm to regimented forest plantations to irrigation canals to livestock marts, are all products of the capitalist exploitation of the countryside.

At the same time, the products of rural capitalism – food, fuel, timber – have often been regarded as too strategically important to be left to the market alone. Hence, the exploitation of the rural has also been sponsored by the state – through the provision of infrastructure, support for research and development, the regulation of markets, and subsidies for producers. Through regulatory regimes such as productivism in post-war Europe, the state colluded with capital in promoting the discourse of the rural as a space of production, as will be discussed further in Chapter 8.

The imperative of capital accumulation requires constant reinvention and innovation, generating new technologies, new crops, new livestock breeds and new products. It has also driven an expansive, dynamic geography, with new sites of production sought in increasingly remote and testing locations, and new markets secured as agri-food commodity chains have become globalized. One consequence is that large areas of countryside in the global north are now at risk of becoming surplus to the requirements of resource capitalism, presenting a political challenge for the state. Should public funds be deployed to subsidize farmers and other producers, to uphold the myth of the production countryside and retain material features such as family farms? Or should non-competitive rural industries be allowed to collapse, potentially taking communities with them?

The concept of multifunctionality, which has gained currency as the productivist paradigm has weakened, suggests that there could be a third way. It contends that productivist and non-productivist activities can coexist in rural regions – even on the same farm – and that both can produce capital. However, in spite of the current popularity of the idea, the principle of coexistence presents both ecological and economic challenges. For example, the risk of contamination from GM crops obstructs the prospect of super-productivism co-exisiting with organic agriculture; whilst the experience of conventionalization in the organic sector suggests that alternative modes of production are vulnerable to appropriation by mainstream agri-industry and the introduction of quasi-productivist methods.

Moreover, if the logic of multifunctionality is extended, then the primacy of production in the rural economy is questioned. By recognizing the full range of functions performed by agriculture (or other exploitative industries), multifunctionality holds that the value of agriculture is not only in the goods that it produces, but also in the social and environmental benefits that it delivers. As an economic strategy, multifunctionality proposes that these non-production benefits should be exploited as a source of revenue, by selling them as commodities to be consumed (for example, through tourism and recreation). As such, multifunctionality begins to engage with another tradition that has long stood as a mirror to the discourse of the rural as a space of production – the idea of the rural as a space of consumption, which the next chapter examines.

FURTHER READING

The development of resource capitalism in late nineteenth- and early twentieth-century California is examined in detail by Richard Walker in *The Conquest of Bread* (2005), which focuses largely on agriculture, and in an article in the *Annals of the Association of American Geographers* (2001) on 'California's golden road to riches: natural resources and regional capitalism, 1848–1940', which focuses more on mining and forestry. George Henderson also interrogates the significance of capital to Californian agriculture in *California and the Fictions of Capital* (1998). For more on the political economy of industrial forestry and its social and environmental impacts, see Scott Prudham's study of Oregon, *Knock on Wood: Nature as Commodity in Douglas-fir Country* (2005). The best introduction to productivism and post-productivism is the chapter by Brian Ilbery and Ian Bowler, 'From agricultural productivism to post-productivism' in *The Geography of Rural Change*, edited by Ilbery (1998), whilst the debates around post-productivism are summarized by Nick Evans, Carol Morris and Michael Winter (2002) in *Progress in Human Geography*. For more on the concept of multifunctionality and its practical application see papers by Geoff Wilson in *Transactions of the Institute of British Geographers* (2001) and the *Journal of Rural Studies* (2008). Julie Guthman's work on organic agriculture in California, published in *Sociologia Ruralis* (2002, 2004), illustrates the evolution of the sector and provides an interesting counterpoint to Walker's study.

4

CONSUMING THE RURAL

INTRODUCTION

For as long as carts have rolled into cities from the countryside laden with crops and fuel and stone, there have been pleasure-seekers who have headed in the other direction, into the country, to hunt, play, stroll, bathe and escape the pressures of urban life. The idea of the rural as a space of production, examined in the previous chapter, has always had a mirror in the similarly powerful idea of the rural as a place of consumption, particularly as a location for leisure and recreation. In some cases, rural sites have simply hosted activities that are not in themselves intrinsically rural – various sports, for example, or, in recent times, theme parks, car boot sales and shopping malls. Commonly, however, the use of rural space for recreation and leisure is tied to an idea of in some way *consuming rurality*, or, at least, consuming attributes associated with an imagined rural idyll.

The consumption of the rural can take many forms. Sightseers visually consume the rural landscape; hill-walkers consume the atmosphere of fresh air and tranquillity; nature-watchers visually consume the wildlife; mountain bikers consume the terrain against which they are pitted; visitors to fayres and festivals consume rural culture; shoppers buy rural crafts, and diners literally consume rural food and drink; and so on. In each of these cases, attributes of rurality that are the object of consumption – scenery, nature, tranquillity, heritage – are translated into commodities that can be bought and sold.

For centuries, the consumption of the rural was a largely (though not exclusively) elite preserve: Roman citizens built villas as rural retreats; mediaeval aristocrats established hunting parks; eighteenth-century gentry created pleasure gardens on their country estates, and embarked on 'grand tours' to consume the scenic treasures of the Alps or the Scottish Highlands. The twentieth century, however, saw the rise of mass consumption culture, tourism and leisure, with travel to rural areas enabled by railways, motor vehicles and commercial flights (the arrival each of which was resisted by opponents who feared that they would disrupt the tranquillity and propriety of the countryside). The regular use of rural space for leisure and recreation by a predominantly urbanized population has transformed public perceptions of the rural in the global north, challenging the political primacy of agriculture. It has also created new opportunities for capital accumulation, and in much of the global north and significant parts of the global south, the consumption economy is now at least as important as the production economy in sustaining rural livelihoods.

This chapter explores the ways in which the rural is consumed, the representations of rurality that are involved, and the material consequences. It begins by introducing the concept of commodification and demonstrating how the practice of rural tourism implicitly involves the commodification of rural landscapes, cultures and experiences. Commodification is critical to the appropriation of rural tourism as a source of capital accumulation, but as the chapter proceeds to discuss, the requirement of capitalism for constant innovation and reinvention can produce the 'creative destruction' of rural consumption sites as landscapes of accumulation are renewed. The remainder of the chapter illustrates the dynamism within rural tourism and recreation by focusing in turn on the different senses engaged in consumption. First, it examines the visual consumption of the rural landscape, framed through the 'rural gaze', and the aural consumption of the rural soundscape. Next, the chapter concentrates on taste, as engaged in food tourism. Finally, the chapter discusses embodied forms of rural consumption that require physical contact with the rural terrain – as practised in various adventure sports – or with nature – as sought through nature tourism.

THE COMMODIFICATION OF THE COUNTRYSIDE

Rural tourism and commodification

The use of rural areas for tourism and recreation has become increasingly widespread across the global north and in many parts of the global south. Although accurate and comprehensive figures for rural tourism are difficult to establish, the evidence that is available points both to the scale of participation in rural tourism and recreational activities and to the sector's value to the rural economy. In England, for example, surveys have found that 39 per cent of the population claim to visit the countryside for pleasure at least once a month, and that the average person makes at least 22 recreational visits to the countryside each year (Defra, 2002; CRC, 2007). Total spending by visitors to rural England has been estimated at UK£12 billion a year – compared with UK£5.6 billion generated by agriculture (Defra, 2007). Indeed, tourism has been widely promoted as a development strategy for rural areas seeking to replace declining primary industries such as agriculture (Cawley and Gillmor, 2008; Storey, 2004).

However, these figures cover a diverse array of leisure activities, including many for which a rural location is largely incidental. Perhaps more usefully, 'rural tourism' might be more narrowly defined as touristic activities that are focused on the consumption of rural landscapes, artefacts, cultures and experiences, involving differing degrees of engagement and performance. At the most passive level, the rural is consumed through 'scenic tourism' (Rojek and Urry, 1997), such as sight-seeing, coach tours, visits to 'beauty spots' and heritage sites, and short walks, where a distance is maintained between the tourist and the rural landscape that is observed. Requiring more active participation are touristic activities that variously involve physical engagement with the rural terrain and environment – from traditional pursuits such as horse-riding and angling to modern adventure sports such as geocaching and jet-boating; encounters with nature, from bird-watching to safaris; or immersion in rural culture – from traditional festivals to working farm holidays. Across this repertoire of activities, rural tourism consistently involves the consumption of rural signifiers as participants seek connection with an imagined idea of the rural. Moreover, tourists are prepared to pay to make this connection – through admission fees and hire charges; expenditure on accommodation, food and drink; the purchase of souvenirs, postcards, local crafts and products, guidebooks and maps; and kitting-out with the 'right' clothing and equipment. As such, rural tourism intrinsically involves the *commodification* of rural signifiers as diverse as 'fresh air' and historic buildings (inputs described by Garrod *et al.* (2006) as 'countryside capital') as products that are in effect bought and sold through tourist consumption.

The concept of commodification (see Box 4.1) provides a theoretical framework for analysing the constitution and impact of rural tourism by

Box 4.1 COMMODIFICATION

An object becomes a commodity when its exchange value – the price that consumers are prepared to pay for the object – exceeds its use value. This transformation is referred to in Marxist political economy theory as commodification. Commodification underpins the market economy because it means that objects are more valuable to producers as goods to be traded than as property to be used. It also recognizes that exchange values can be inflated by cultural fashions and preferences – the exchange value of antique furniture, for example, can far exceed its usefulness for its owners. Furthermore, commodification also takes place when entities that have not traditionally been considered in economic terms are ascribed with an economic value – in a rural context this can include things such as landscape, fresh air, nature observation and adventures, or, equally, biodiversity and carbon sequestration.

As capitalism has advanced, an increasingly extensive array of objects, functions and experiences have been transformed into commodities, and the exchange values ascribed to commodities have increasingly been informed by cultural perceptions and have become distanced from the commodity's objective usefulness and its conditions of production. Steven Best (1989) theorizes this as a progression from the *society of the commodity*, described by Karl Marx, to the *society of the spectacle*, and subsequently to the *society of the simulacrum*. The idea of the society of the spectacle derives from the work of Guy Debord that holds that social control is maintained by the mass consumption of a world of spectacles created

continued

by others and designed to pacify and depoliticize. This is a world dominated by mass media, entertainment, leisure and tourism, and involves the extension of commodification to previously non-commodified areas of life, with commodities that are frequently abstract or virtual and aimed at providing experiences.

In the society of the simulacrum commodification takes a further step into the hyper-real. Taken from Jean Baudrillard's description of post-modernity, the society of the simulacrum is based on consumption and information technology. As Cloke (1993) explains, in this society 'exchange is carried out at the level of signs, images and information, and commodification is not just about the selling of an object in terms of its image and spectacle, it is in an abstract sense the absorption of the object into the image so as to allow exchange to take place in semiotic form' (p. 56). Consumption hence becomes organized around 'signs' and representative commodities that have no form or basis in reality. Arguably such hyper-real commodities have no use value, only an exchange value.

Further reading: Best (1989).

drawing attention first to the ways in which the rural is 'packaged' and 'sold' to consumers, and second to the material effects of the tourist economy in the rural landscape. As commentaries including Cloke (1993), Crouch (2006) and Perkins (2006) have described, the commodification of the countryside for touristic consumption involves the enrolment and repackaging of a wide range of rural objects and experiences, responding to a range of different expectations of the rural. In some cases, objects that already exist as commodities are ascribed with new meanings, and new value, as rural signifiers: food products, for example, are given enhanced value when re-presented as regional specialties; whilst vintage tractors are cherished as icons of agrarian heritage. In other cases, activities that have traditionally been part of everyday recreation in the countryside, such as boating, cycling, horse-riding, fishing and walking, are transformed into experiences to be sold to tourists (Perkins, 2006), sometimes enhanced with added technology:

> Jet boating, rafting and kayaking on white water, four-wheel drive vehicle adventures, mountain biking, some ecotourism activities and walking using sophisticated back country lodges are all examples of technologically based commodified rural recreational activities catering for local and international visitors on a fee-paying basis.
>
> (Perkins, 2006: 252–53)

More controversially, commodification also extends to more abstract components of the rural experience, such as scenic views, fresh air and tranquillity, which are translated into objects of consumption for profit. The capitalization of these elements may be achieved directly – for example, through pay-to-enter viewpoints – or indirectly – for example, through their capture in materials marketing rural places to prospective tourists.

Indeed, marketing is a critical component in the commodification of the countryside. The creation of new meaning through commodification is a collaborative process involving both the producer and the consumer (Bell, 2006; Crouch, 2006; Perkins, 2006). To be attractive, the commodified rural must correspond with the expectations of the rural carried by consumers, which are shaped by various cultural influences (see Chapter 2). Yet, popular perceptions of the countryside are also informed by the representations employed to promote rural places as tourist destinations, which selectively emphasize particular signifiers of rurality. Cloke (1993), for example, in an analysis of brochures for tourist attractions in rural Britain, notes the repeated evocation of landscape, nature, history, the opportunity to purchase crafts and country fayre, and family safety, even for theme-park type attractions.

As such, the representations of the commodified countryside deployed in marketing to tourists become detached from the material realities of the places concerned. Perkins (2006), for example, cites an analysis by Hughes (1992) of promotional material for the Scottish Tourist Board which 'created a sanitized and picturesque mythical Highlands quite unlike the real poverty-striken and oppressed Highlands of history' (Perkins, 2006: 252). A similar critique can be advanced of the marketing of rural tourist destinations in the global south in which exotic landscapes and wildlife are emphasized, but problems of extreme poverty and political oppression ignored.

It is in this way that the commodification of the countryside proceeds along the sequence outlined by Best (1989) from the society of the commodity to the society of the simulacrum (Perkins, 2006; see also Box 4.1). The society of the commodity was established in rural areas long before the rise of mass rural tourism by resource capitalism which produced goods whose exchange value exceeded their direct usefulness to rural residents (see Chapter 3). The early incursion of the consumption economy simply created new opportunities for enhanced exchange values to be attached to objects whose use value for production had declined. For example, farms started to obtain an additional income by converting disused outbuildings into holiday accommodation, or turning fields into paying campsites. However, the rise of mass rural tourism also coincided with the emergence of the society of the spectacle. By offering spectacles of landscape, nature and cultural performance, and by reinforcing motifs of tranquillity, order, tradition and security, the commodified countryside can be seen as making a major contribution to the function of spectacle in pacifying and depoliticizing society.

The transition to the society of the simulacrum is marked, as Perkins (2006) observes, by 'a further step into hyper-reality and has as its focus the consumption of representative commodities which have no form or basis in reality' (p. 247). The society of the simulacrum hence implies the consumption of rural signifiers that have become wholly detached from a materially embedded rurality and exist purely as a virtual or hyper-rural designed exclusively for consumption. Examples might include fantasy landscapes inscribed with the place-names and narratives of film, television programmes or literature (Mordue, 1999; Squire, 1992); staged performances of farm work, craft work or rural 'traditions' (Edensor, 2006); rural crafts and artefacts produced solely for the tourist market; and various theme parks, farm parks, shopping villages and holiday villages that offer stylized and exaggerated reproductions of rural landscapes for paying visitors (Mitchell, 1998; Wilson, 1992).

Place, commodification and creative destruction

The commodification of the countryside transforms those rural places that become the theatres of consumption for tourists and visitors. The changes can start with speculative ventures, including proactive initiatives by entrepreneurs, local governments and business associations such as the creation of nature parks or town trails, or the conversion of farm buildings or historical sites into tourist attractions. As visitor numbers increase, more businesses will be established to cater to tourists, and the rural landscape itself may be modified to meet tourist expectations, as might other aspects of rural life such as local festivals. However, increased visitor numbers can also have a negative impact, through congestion and pollution and the development of commercial ventures that are more and more removed from the original rural attractions – fast food restaurants, for example, or nightclubs. Eventually, a point may be reached where visitor numbers start to fall because the locality can no longer offer the 'rural' experience that prompted the growth of tourism in the first place.

Mitchell (1998) describes this cycle as **creative destruction**, borrowing a concept initially developed by David Harvey (1985) to refer to the creation and destruction of products that it is intrinsic to a capitalist cycle of accumulation – for example, as technological innovations make previous products obsolete. Mitchell adapts the concept for rural tourism and consumption, identifying five stages in the cycle of creative destruction in commodified rural landscapes: early commodification; advanced commodification; early destruction; advanced destruction; and post-destruction.

In the first stage of early commodification, tourists and other consumers are attracted to rural sites presenting opportunities for entrepreneurs. Proceeds are then re-invested, and sites attract both additional investors and additional consumers, fuelling advanced commodification in which commodities are increasingly detached from their original rural referents. However, growing visitor numbers, commercialization, and their associated impacts of noise, congestion, litter and pollution, start to detract from rural attributes (such as tranquillity or authenticity) that initially attracted consumers, prompting the early stages of destruction. As these negative factors intensify, the destruction of the 'product' becomes more advanced as consumption levels fall. In the final post-destruction stage, consumers and investors have switched their attention elsewhere and once commodified landscapes are either abandoned or become sites for new investment, which may not necessarily be linked to tourism or consumption (Mitchell, 1998; Mitchell and de Waal, 2009).

The community of St Jacobs in southern Ontario, Canada, which Mitchell studied, provides a good illustration of this process. The small town of around 1,500 people had historically functioned as a service centre for the Old Order Mennonite community, a distinctive cultural and religious group that rejects modern technology and continues to hold to traditions such as horse-and-trap transport. Curiosity about the Mennonites started to attract visitors from neighbouring cities including Waterloo and Toronto during the 1970s, leading a local entrepreneur to open a Mennonite-themed gift shop and restaurant, quickly followed by the conversion of a decommissioned flour mill into a saleroom for reproduction pine furniture, and the establishment of a village bakery. As tourist numbers increased, further businesses were opened on a similar theme, turning St Jacobs into a heritage shopping village, with general support from local residents (Mitchell, 1998).

Between 1979 and 1990, at least C$8.4 million was invested in St Jacobs by the main developer, supplemented by other investors who opened shops and attractions selling versions of the area's rural heritage and tradition to consumers. By 1989, a million people a year were visiting the town. In the stage of advanced commodification the whole town began to be marketed in terms of heritage consumption, with tourists encouraged to visit 'the village where time has stood still' (slogan, quoted by Mitchell, 1998: 280).

By 1995, there were over one hundred businesses in St Jacobs, predominantly linked to tourism, and commercial ventures had spread beyond the historic core to new edge-of-town developments. However, Mitchell (1998) observes that the scale of commodification had 'forced the Mennonite population to seek out less contrived landscapes for their shopping needs' and 'caused partial destruction of the rural idyll in the eyes of some village residents' (p. 281), marking a transition to the stage of early destruction. Investment in St Jacobs has continued since 1995, including by national and international chains, increasing visitor numbers further, but also changing the character of the retail mix (Mitchell and de Waal, 2009). Discontent from local residents has also continued, with some moving out of the town; but, significantly, some of the early businesses have also started to leave. Mitchell and de Waal give the example of a store selling traditional 'smocking' embroidery, which closed in 2000 because the 'touristy location no longer felt right' (p. 159).

Creative destruction is not restricted to locations in the global north. Fan *et al.* (2008) apply the model to the 'water town' of Luzhi in China, an ancient river-based settlement that attracts tourists both for its history and for its idyllic setting. In the Chinese political-economic system, the state has played a key role in promoting development in Luzhi, but tourist numbers and private enterprises serving tourists have expanded with liberal reforms since the mid-1990s. Yet, Fan *et al.* (2008) argue that whilst clear economic benefits have been realized, there is growing awareness among local residents of the negative impacts of tourism and commodification, corresponding with the model of creative destruction.

SIGHTS, SOUNDS AND TASTES OF THE COUNTRYSIDE

Scenic tourism

The touristic consumption of the countryside is a multi-sensory experience, but the first and most important dimension is visual consumption (Plate 4.1). The vogue for 'sight-seeing' emerged in the eighteenth century and rapidly supplanted the improvement of the mind, or the treatment of the body (at spas), as the main reason for leisure travel for those who could afford it (Alder, 1989). The development of railways in the

Plate 4.1 Sight-seeing in the Rocky Mountain National Park, Colorado (Photo: author)

nineteenth century revolutionized leisure travel, granting easy access to rural areas for the masses and opening up landscapes whose perceived 'scenic' quality often reflected their isolation and the lack of urbanization (Bunce 1994). Entrepreneurs in these regions were quick to spot the opportunity to profit from the new fashion, promoting new rural resorts with striking visual depictions of landscapes that had to be seen.

One such location was Lake Memphremagog on the border of Canada and the United States, between Quebec and Vermont. The impressive scenery of the 27 mile long lake, surrounded by forests and mountains, had been celebrated by travellers even before the arrival of the Connecticut and Passumpsic Railroad in 1864, but it was the railway that brought mass tourism and prompted the development of hotels and other tourist facilities and attractions around the lake. As Little (2009) records, 'with the aid of outside capital, [entrepreneurs] built steamboats and resort hotels, relying upon guidebooks and newspaper correspondents to present an image of the lake as an unspoilt pictureseque wilderness at a time when the "fashionable" resorts such as Saratoga Springs were increasingly criticized as places of artifice and immorality' (pp. 720–21).

The introduction of steamboat trips on the lake provided visitors with a means of viewing the scenery around the lake from a unique perspective, pushing beyond the limits of the roads and railways into the 'wilderness'. Little (2009) notes that the long, narrow shape of the lake meant that 'the view was described in the promotional literature as a slowly unfolding panorama', whilst the slow pace of the steamers encouraged passengers 'to develop a spiritual affinity with their scenic surroundings' (p. 717). Although pursuits such as hiking and angling began to gain popularity in the area around Lake Memphremagog, it was clear that the prevailing attraction for most tourists was the scenery. Even the climb to the local pinnacle of Owl's Head was undertaken primarily for the view from the top, described as combining 'the grand and the beautiful – sweeping vistas and "a few gem-like islands 'well set' and luxuriantly covered"' (Little, 2009: 724). The rural landscape was hence something to be consumed visually from an appropriate distance, and was commodified through the steamboats, walking trails and hotel terraces that provided visitors with a 'view'.

The rural gaze

The act of viewing the rural landscape, however, is not as simple or as neutral as it might initially seem. As Abram (2003) writes, 'looking is the active organization of what we see, and what we see is socially organized, structured through our internal interpretation of the visual stimulus' (p. 31). This organized and systematic 'way of seeing' was conceptualized by Foucault (1976) as the 'gaze' – an act of power in which collective social norms define not only how we interpret the things that we see, but also what we actually see (and do not see) and where we look. Foucault was primarily interested in the 'medical gaze', demonstrating, for example, that until the eighteenth century mental illness was not seen, because the symptoms of mental illness were not understood by society as related to health. Observers thus saw people who were 'possessed' not people with mental illness.

Urry (1990) drew on Foucault to propose the notion of the *tourist gaze*. This contends that the way in which a tourist sees landscape, or cultural events, or other objects of their visual consumption, is socially conditioned. Our evaluation of what is scenic, and what is not, our appreciation of the aesthetics of landscape, and our judgements about authenticity and naturalness, are all shaped by social and cultural norms and influences, by our education, by the prompting of guidebooks and brochures, and by the images that we have consumed through film, television, art, photography and other media. The tourist gaze also informs our selective viewing of the landscape. It means that we often do not see the people who live and work in the landscape, the labour that goes into maintaining a landscape, or the poverty that hides behind the door of a picturesque but run-down cottage.

The idea of the gaze is further extended by Abram (2003), who introduces the *rural gaze* as a mechanism for exploring how the ways in which people see the countryside are similarly socially constructed. Abram acknowledges that the rural gaze and the tourist gaze converge in framing a nostalgic vision of the countryside for touristic consumption, noting that 'the tourist gaze upon the rural landscape is one and the same as the rural gaze which aestheticizes land uses in a nostalgic way in an attempt to distance it from contemporary capital and globalizing processes' (p. 35). However, she argues that the rural gaze extends beyond tourism and is evident, for example, in the motivations and responses of in-migrants to rural communities, in attitudes to conservation and preservation, in land use planning policies and development control, and in land management decisions, as well as in obscuring the recognition of problems such as poverty and deprivation in rural areas (see Box 4.2).

The conditioning of the tourist gaze over the rural landscape has been in part the result of deliberate attempts to direct the observer's view,

Box 4.2 BUYING THE VIEW – MIGRATION AND GENTRIFICATION

The consumption of the countryside includes not only the growth of rural tourism and recreation, but also the purchase of rural properties, either as holiday homes or as new permanent residences, by urban dwellers. There are many reasons why people move to rural areas, with migration for employment and family reasons being the most common. However, studies in a number of countries, including Australia, Britain, the Netherlands, Norway and the United States, have identified the desire to consume rurality as a key factor in migration decisions (Burnley and Murphy, 2004; Flognfeldt, 2006; Halfacree, 1994; Nelson, 2006; van Dam *et al.*, 2002). The anticipated consumption crosses the range of sensory experiences discussed in this chapter. For some migrants the major attraction is the prospect of participating in rural community life, or assuming a rural lifestyle (discussed further in Chapter 6). For others, moving to the country makes it easier to enjoy the landscape in recreational activities such as walking and cycling.

The selection of specific properties for purchase by individual in-migrants is frequently informed by preferences for visual or aural consumption. Houses that command views of bucolic rural landscapes (or mountain or coastal scenery) are highly cherished, with prices inflated accordingly. Villages and small towns that conform visually to the expectations of the rural gaze, with old buildings constructed in the local vernacular and rustic features such as village greens or historic stone crosses, are similarly favoured over settlements characterized by modern, standardized buildings (Abram, 2003). Tranquillity can also be a highly sought-after quality, especially by investors in more remote properties (Smith and Phillips, 2001). Phillips (2002), for example, quotes an in-migrant to a rural community in Berkshire, southern England, explaining that, 'I chose this village to retire to particularly for its peace, tranquillity, beauty and rural aspect in an otherwise rather noisy world' (p. 300).

The process of buying a house or other property in the country is in itself an act of consumption, and is part of the commodification of the rural. House prices in many sought-after rural areas of Europe, North America, Australia and New Zealand have soared over the last two decades in response to demand from would-be in-migrants, and at an individual scale reflect value being placed on attributes such as historic character, view and tranquillity. The inflation of house prices in these areas has frequently reduced the affordability of property for residents earning average local wages, thus contributing to the social recomposition of rural communities as established local families are replaced by wealthier ex-urban middle-class in-migrants, in a process known as gentrification (Smith and Phillips, 2001; Phillips, 2002). Studies of urban gentrification have described the process as a 'consumption-biased spatial complex' (Zukin, 1990), noting that 'residential gentrification has often been accompanied, and indeed arguably stimulated, by the development of retail, leisure and entertainment facilities such as restaurants, bars, clubs, fashion boutiques, art galleries, museums and sports facilities' (Phillips, 2002: 286).

Phillips (2002) argues that whilst rural gentrification seemingly takes a different form, it can also be described as a consumption-biased spatial complex that acts as a focus for investment of labour power and financial capital, producing a symbolic product that contributes to the further circulation of capital. He observes, for example, 'the rapid gains which can be made from buying and selling houses in the countryside, and also the existence of a significant number of rural residents who were undertaking substantial building work on their property which acted to heighten its exchange value' (p. 286). Equally, Phillips points to the transformation of redundant rural buildings such as former schools, railway stations, chapels and barns into retail and leisure facilities serving the lifestyle demands of middle-class residents and visitors; and to the evoking of a middle-class rural aesthetic in advertisements for products such as cars and fitted kitchens.

Further reading: Phillips (2002), Smith and Phillips (2001).

in some cases for commercial purposes, in other cases from an elite belief that the masses needed to be taught how to see the countryside. The perceived need for visual education was stoked in particular by the rise of the motor car during the inter-war years. Not only did motorized transport enable more people to flow more freely through the countryside, exploring rural back-lanes far away from the constraints of the rail network, but it also introduced a new way of seeing the rural landscape. Whereas rail travel restricts the view of the passenger (Schivelbusch, 1986), motor transport created a new perspective: the view from the road. Bus and coach operators initiated sight-seeing trips, with one company noting that its new-fangled charabancs enabled 'the onlooker to see and appreciate all that is delectable in the countryside' (quoted by Brace, 2003: 53).

The car afforded even more flexibility and soon an infrastructure began to emerge that was designed to support car-based sight-seeing and to direct the gaze of motorists and their passengers. This included the creation of scenic roads, such as the Blue Ridge Parkway that runs through the Appalachian Mountains of Virginia and North Carolina in the eastern United States. Built during the New Deal era, the Blue Ridge Parkway was designed expressly for leisure traffic. The route rewarded motorists with panoramas over striking natural scenery, but the project also involved the modification of the road's surroundings 'to create a landscape pleasing to the motorist, which involved using the land in a way that would "make an attractive picture from the Parkway"' (Wilson, 1992: 35). Moreover, motorists were guided through the landscape by roadside signage that was both informative and interpretative:

> like railroads, the Parkway is periodically marked by mileposts, their purpose being to orient motorists vis-à-vis their itineraries and to aid road maintenance and administration. Talked about in the original plans as a way of relieving monotony, the mileposts also introduce the notion of progress to the motorist's experience of the landscape; the miles tick off as nature unfolds magnificently before us. The Parkway has a logo – a circle enclosing a roadway, a mountain peak and a wind-swept white pine – and like all logos it is repeated. Other road signage, especially at the entrances, is standardized to underline the special quality of this created environment. Gouged wood signs point out road elevations, local history, and the names of distant

> features of the landscape. Other diversions organize the motor tour: parking overlooks, short hiking trails, local museums, campgrounds, and parks space every thirty miles. In this way, the planners designed tourist movement into the land itself.
>
> (Wilson, 1992: 35)

Motorists and their passengers could also turn to other sources of guidance. Maps, for instance, acquired a new purpose and some started to mark scenic roads and viewpoints and other features of interest to the tourist. Guidebooks also proliferated, such as the British Heritage series published by Batsford. Although restricted to line-drawings and engravings for images, the text of these guides nonetheless conveyed a vivid picture of the British countryside that reproduced the nostalgic discourse of the rural gaze. In doing so, they shaped the way of seeing the rural landscape for thousands of tourists.

Some volumes were more overtly instructional, including *How to See the Countryside*, by Harry Batsford, published in 1940 (Brace, 2003), and *The Countryside Companion* by Tom Stephenson, published in 1939, which contained a chapter with the same title (Matless, 1998). These texts 'argued that while landscape can be appreciated without being understood, it was best appreciated given an understanding of form and structure' (Brace, 2003: 58), and as such set out to educate readers about nature study, landscape history, architecture and map-reading. They formed part of a broader disciplining of rural leisure that was both cultural and spatial. Tourists were told how to act and behave in the countryside – for example, through the advisory rules of the 'Country Code' adopted in England and Wales in 1951 (Merriman, 2005; Parker, 2006) – but also directed towards particular spaces of rural consumption, such as national parks.

The direction of the tourist's rural gaze may be less didactic today, but it is no less prevalent. Guidebooks with glossy photographs, publicity brochures, postcards, visitor centre displays, walking leaflets and trail markers and information boards tell us what to see and how to interpret the scenery, whilst signposted viewpoints, orientation boards and even carefully positioned benches direct us on where to look. These prompts form part of the commodification of the rural landscape for visual consumption, and even where they do not directly involve a charge or payment, they tend to lead tourists to tea-rooms or fee-paying car parks or gift shops, where the same view that the tourist has witnessed can be found reproduced across an array of souvenirs. As Macnaghten and Urry (1998) observe, 'policies which focus on the beauty of the countryside as opposed to one's freedom over it clearly invite new methods of visitor management, and policies of visitor management can easily lead to new methods of payment for use' (p. 192).

However, the rural gaze is not formed solely by the directing of guidebooks and signposts, but is also constituted by deeper cultural experiences and collective knowledge. Landscape painting existed as a form of commodification of the rural landscape long before the advent of the motor car, and its conventions have strongly informed the way in which people have viewed and interpreted actual rural landscapes. Not only have tourists sought out the panoramas depicted in their favourite paintings, but they have also looked for the same aesthetic qualities appreciated in art in the material landscape. Film and television have similarly come to play an important role in shaping the rural gaze (see also Chapter 2), with tourists seeking out the landscapes portrayed – for example, in tours to filming locations for the *Lords of the Rings* triology in New Zealand (Tzanelli, 2004). As tourists expect to see the landscapes exactly as seen on screen, cinematic tourism can demand the blurring of fictional and 'real' landscapes, especially in the interpretative notes that guide the visitor around a site (Plate 4.2).

Yet, the 'reality' of any landscape can be questioned. Marxist art critics describe landscape as a 'visual ideology' (Cosgrove, 1985: 47) that has been employed by a ruling elite to legitimize their position and to 'mystify' the power-relations that underpin it (see also Berger, 1972; Wylie, 2007). Property ownership and 'improvement' is celebrated, whilst the labour that sustains agriculture, for example, and the social inequalities of the countryside are disguised or omitted. As such it aestheticizes and neutralizes the operation of rural power. Similarly, Daniels (1989) attacked the duplicity of landscape with respect to human relations with nature, as Wylie observes:

> Daniels argues that landscape, as a way of seeing, is duplicitous, because whilst on the one hand it offers a redemptive, transcendent and aesthetic vision of sensual unity with nature, on the other it operates as a smokescreen concealing the underlying truth of material conditions and manipulating our vision such that we have become unaware of the distancing that separates us from the natural world.
>
> (Wylie, 2007: 67)

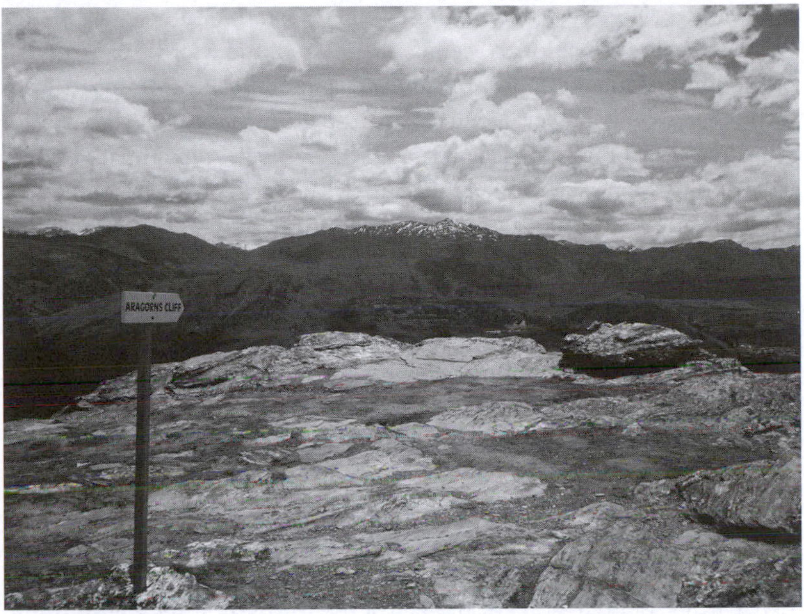

Plate 4.2 *Lord of the Rings* film tourism, Queenstown, New Zealand (Photo: author)

Whilst these critiques are aimed primarily at landscape art, the challenge to 'landscape' as a way of seeing makes clear that they also extend to ways of viewing the rural that reduce complex rural environments to the rather two-dimensional and simplifying notion of 'landscape'.

The rural gaze is hence a political act, and conflicts can break to the surface if the expectations of the gaze clash with economic or other imperatives driving change in rural areas. For example, agricultural practices that involve the removal of hedgerows, the ploughing of grassland or changes in crop type, thus altering the appearance of the rural landscape, can provoke conflicts, as can deforestation, afforestation and the construction of new roads or new housing. Opposition to windfarms is frequently driven not by environmental concerns, but by anger at the visual impact, the spoiling of a view (Woods, 2003b). In fact, campaigns for protection for areas of scenic countryside extend back to the nineteenth century, and were a key factor in the introduction of the town and country planning system in Britain (Bunce, 1994; Matless, 1998). Many national parks and other protected areas were designated not for environmental reasons but to preserve a visual amenity. Scotland, for example, has 40 'National Scenic Areas', and there are 50 'Areas of Outstanding Natural Beauty' in England and Wales. The designation of such areas is not, however, purely altruistic. Protected landscapes safeguard the visual capital that is represented by their scenic qualities, and act as additional markers attracting tourists and facilitating the commodification of the landscape for economic gain.

The ubiquity of visual consumption as a way of engaging the countryside is such that it can be difficult to comprehend the idea of consuming the rural landscape without sight. MacPherson (2009) describes her experience of ethnographic fieldwork with a group of visually impaired walkers and their sighted guides in England. As she reports, 'far from rendering sight-based practices of landscape apprehension redundant, experiences of visual impairment bring both symbolic and perceptual issues of sight and landscape to the fore' (p. 1044). MacPherson presents the consumption of the rural landscape as an intercorporeal experience, negotiated between bodies as the sighted guide describes the landscape to their partner (and 'negotiates the landscape for two' in moving through gates etc.), with the visually impaired walker combining the described view with their residual vision, their visual memory and their imagination to see the scenery in their 'mind's eye'. She notes that the sighted guides select the 'classic scenic trails' of areas such as England's Lake District for walks, and that the landscape as 'seen' and understood by the visually impaired walkers is replete with the same romanticized, pastoral iconography as the sighted rural gaze. At the same time, she notes the pleasure that the walkers described feeling at the response of their sighted guides to scenery, conveyed through tone of voice and touch.

Thus, whilst MacPherson (2009) does not directly discuss the role of other senses, such as sound, taste, smell and touch, to the visually impaired walkers' consumption of the rural landscape, the intercorporeal negotiation of the landscape that she describes is in itself a multi-sensory experience and points to the function that other senses play in the consumption of the countryside, beyond the rural gaze.

Rural soundscapes

The visual consumption of rural landscapes is strongly associated with the aural consumption of rural soundscapes. As Matsinos *et al.* (2008) state, 'the origin and intensity of ... acoustic signals reflect the structure and spatial configuration of the landscape since human activities, biological processes and natural phenomena produce sounds which act as "messengers" of the landscape' (p. 946). The reception and interpretation of sounds is therefore part of the way in which we 'see' and understand landscape, and our socially conditioned response to different sounds and noises, heard in particular places, adds another dimension to the 'rural gaze'.

Ideas of the rural idyll, for example, are primarily represented visually (the *image* of rolling farmland, or of the lake in the forest), but imply a soundscape through the use of terms such as 'peaceful', 'quiet', 'tranquil'. The idealized rural soundscape is hence commonly constructed in terms of the absence of urban noises, such as traffic, machinery, construction work, loud music and Tannoy announcements. This idealized soundscape can be summed up by the concept of tranquillity, which refers to a soothing, calming environment. The absence of disruptive and stressful noise is central to the notion of tranquillity, although sight and sound can be intimately entwined in the sensation of tranquillity. For instance, Herzog and Barnes (1999), in a psychological experiment that measured perceptions of tranquillity associated with different landscapes, defined tranquillity as 'how much ... you think this setting is a *quiet*, peaceful place, a good place to get away from the demands of everyday life' (p. 173; my emphasis), but assessed responses solely to the visual stimulus of photographs.

The association of rurality and tranquillity is indicated by the preference expressed by Herzog and Barnes's respondents for pastoral landscapes, and reinforces the discourse of the countryside as a place of escape from the city. As such, the protection of tranquillity has long been a key objective of rural preservationist groups alongside the protection of the visual appearance of rural landscapes. The Campaign to Protect Rural England (CPRE), for example, argues that:

> Tranquillity is important for everyone – for our hearts, minds and bodies. We all need to 'get away from it all' every now and then. Tranquil areas provide a means of doing this in a crowded, heavily built-up country. Being largely natural and free from intrusive man-made noise and structures, tranquil areas in the countryside allow us to escape the noise and stress of cities, towns and suburbs, to be inspired and to get refreshed.
>
> (CPRE website)

In support of its campaign, the CPRE has pioneered the production of 'tranquillity maps', which purport to identify the 'tranquil areas' of England. The significance of sound is immediately evident in this cartography, with 'non-tranquil' zones around noise generators such as major roads, airports and military airfields intruding into the general tranquillity of the rural areas. The primary purpose of the maps is to highlight the gradual erosion of tranquil areas over time with the spread of urbanization and increased road and air traffic, but they also contribute to a social conditioning parallel to that of the rural gaze, in which a moral geography of sound in the countryside is constructed and reproduced.

Tranquillity is not, however, the absence of noise. The CPRE website invites visitors to 'hear a piece of tranquillity', with audio clips of various rural settings in which birdsong features prominently, but also streams and waterfalls, farmyard sounds and weather such as wind and thunder. These are the sounds that constitute the idyllic rendering of the rural soundscape, along with the noise of traditional rural crafts or activities such as black-smithing, or church bell-ringing, and some forms of music (see Box 4.3). Threats to the idyllized rural soundscape hence come not only from the intrusion of 'urban' noise such as traffic and machinery, but also from modern farming techniques that have reduced populations of farmland birds, and from cultural change that suppresses rural traditions. Locations that can offer these traditional, natural soundscapes without the disruption of urban noise become commodified as sites for the aural consumption of the countryside by tourists seeking tranquillity. Yet, the commodification of such sites is precarious, and can easily tip into Mitchell's cycle of creative destruction as increasing visitor numbers generate traffic and other noise that impairs the sought-after tranquil soundscape.

The struggle to control the rural soundscape is illustrated by Matless (2005) in a case study of the Norfolk Broads wetlands in England. The region is renowned for its rich natural environment and wildlife, which visitors experience in part as a unique soundscape, as descriptions quoted by Matless convey:

> A warm summer night has a special music. The thing to listen for is the singing water-weed. You almost have to hold your breath to hear it ... It's a queer experience; a sort of fairy music rises from the water. If your ear is not attuned you might compare it with the 'muzz' in sound broadcasting before VHF came along; but it's really rather like

Box 4.3 MUSIC IN THE RURAL SOUNDSCAPE

An important component in the rural soundscape for some residents and visitors alike is the playing of genres of music that are associated with rurality, notably folk, country and bluegrass music, especially where the music is evocative of a particular place or region. Performances of folk or country music can be ways of performing rurality as a way of life or identity, particularly where the lyrics relate distinctively rural experiences and concerns, as is discussed further in Chapter 7. However, listening to music can also be a way of consuming rurality. Rural music has been commodified for mass consumption through recordings and the sale of CDs and albums worldwide; but commodification also occurs through the staging of live music at gigs and festivals in rural settings, and the associated growth of music tourism.

The town of Tamworth, in New South Wales, Australia, is a prime example of music-based commodification (Gibson and Davidson, 2004). In the 1970s, a local radio station established an annual festival in the town for Australian country music. At first, the festival was just one of several country music festivals hosted by rural towns around Australia as mostly local events, but it expanded rapidly, becoming a major commercial event, sustained through an actor-network including the Australian music industry, local and regional government, national and local media, sponsors including international brands, as well as the promoters and performers from Australia and abroad. By the start of the twenty-first century, the festival involved a ten-day programme with 2,400 events, 116 venues, nearly 1,000 artists and around 60,000 visitors to the town (Gibson and Davidson, 2004), making a major contribution to the local economy. Moreover, the festival had become the centrepiece for the broader re-branding of Tamworth as 'Australia's Country Music Capital', with the aim of attracting tourists beyond the festival period.

In the process of commodification, the rural identity of Tamworth and the rural character of country music have become mutually reinforcing. Gibson and Davidson (2004) observe that,

continued

'the organizers claim that 'country music conjures up all things rural', a sentiment that can be brought back from the festival in paintings of rural landscapes captured on anything from a beer bottle to a garden spade' (p. 395). As such, they argue, 'Tamworth is depicted as the site of quintessential rural experiences' (*Ibid*), with the town discursively positioned at the centre of an 'Australian nationalism constructed through rurality' (*Ibid*).

Not all live music consumed in the countryside makes claims to be characteristically rural. Rural locations have become favoured settings for large-scale concerts and festivals embracing a diversity of genres, from classical music and opera (such as the Glyndebourne festival in England), to rock and pop, such as the Glastonbury Festival, held on a working dairy farm in south-west England, that attracts over 175,000 people each year. Some festivals trade on their rural setting, presenting themselves as relaxed tasters of rural life, but others are more expedient attempts by rural towns to occupy novel niches in the competitive market for rural tourism. The small town of Parkes in New South Wales, for example, hosts an annual Elvis Revival Festival for Elvis tribute acts, despite having no direct connection with Elvis Presley or his music (Brennan-Horley *et al.*, 2007). In economic terms, however, the event has proved highly successful, with Brennan-Horley *et al.* (2007), calculating that over A$1.1 million is spent in the town over the weekend.

Music tourism in rural areas can be controversial, however. Large concerts and festivals, especially for rock and pop music, have a sonic impact which is diametrically opposed to the idea of tranquillity, and many face hostility from local residents. Even festivals that celebrate endogenous rural music can create conflict over the associated congestion and the way in which the local culture gets represented, although Gibson and Davidson (2004) report that local opposition to the Tamworth Country Music Festival waned as the contribution to the local economy became evident.

Further reading: Gibson and Davidson (2004).

> a faint jifflin of cymbals, or the hiss of rain on water. Only one sort of weed performs in this way; the hornwort.
>
> (Radio talk by naturalist Ted Ellis, 1957, quoted by Matless, 2005: 750–51)

> Perhaps the most characteristic of the Broadland birds is that weird member of the heron tribe – the bittern. Its eerie foghorn booming, the shrill whistle of otters and the pig-like grunting of water-rails are characteristic sounds of the Broadland night.
>
> (National Parks Committee report, 1947, quoted by Matless, 2005: 751)

The prospect of hearing a bittern, or a water-rail or the hornwort, has been part of the draw for tourists to the Broads since the nineteenth century, but the commodification of the region's waterways for tourism also introduced less welcome noise. The shouts, chatter and singing of tourists and the playing of 'gramophone' music and radios from pleasure boats formed a preoccupation for writers on the region in the early twentieth century, as Matless documents. One early guidebook instructed visitors: 'Don't play the piano in season and out of season (the reedbird's song is sweeter on the Broads)' (Matless, 2005: 753), and cautioned that 'sound travels a long way on the water' (*Ibid*: 754). The authorities responsible for governing the district, meanwhile, debated strategies for regulating the transgressive mobility of 'inappropriate sound' (*Ibid*: 753). However, as Matless observes, 'if sounds are deemed out of place this is not for intruding into silence but from disrupting an acoustic ecology whose "silence" is already full of sounds, some non-human – bird song, reed and water – some the product of human activity – sailing, land work' (p. 760).

The search for tranquillity has pushed tourists into more and more remote and exotic rural settings, and for some at least the sense of a 'quest' is heightened by spiritual motivations. There is a long history of religious retreats to the countryside for quiet contemplation, but the tradition has been secularized with the commodification and consumption of a 'wilderness spirituality' in places such as the Tasmanian Wilderness World Heritage Area in Australia (Ashley, 2007). The Tasmanian Parks and Wildlife Service notes that 'wilderness areas offer a place for relaxation, reflection and spiritual renewal' (Ashley, 2007: 61), with the latter achieved through a communing with nature that has visual, haptic and sonic dimensions. Spiritual renewal is hence portrayed as a 'quiet' practice (see also Holloway, 2003, on spiritual seekers on Glastonbury Tor in England), with the tranquillity of rural settings associated with a lack of distractions, thus enabling an openness of spiritual experience.

Tasting the country

In the same way that scenic tourism has supported the commodification of rural landscapes that have lost value as sites for productivist agriculture, opportunities for the re-commodification of traditional food and drink products, and of traditional systems of production and consumption of food and drink, have been exploited by the growth of food tourism. Whereas scenic tourism involves the visual consumption of the countryside, food tourism seeks to consume rural through taste. Bell (2006) identifies this as the pursuit of the 'gastro-idyll', a nostalgic rendering of the rural as a place in which good, wholesome, fresh food can be eaten according to regional traditions and recipes in convivial surroundings. As such, food tourism is part of a broader re-connection of taste and place, reversing the separation of the spaces of production and of consumption of food that had been a feature of productivism (Gouveia and Juska, 2002; Tregear, 2003). This movement has promoted the sale and consumption of local food in rural areas – for example, through farmer's markets – as well as the rediscovery of regional recipes and culinary traditions (Hinrichs, 2003; Holloway and Kneafsey, 2000; Kimura and Nishiyama, 2008; Miele and Murdoch, 2002). At the same time, distinctive regional foodstuffs can command a higher price as 'authentic' rural produce, as indicated by schemes such as Protected Designations of Origin (PDO) and Protected Geographical Indications (PGI) in the European Union (Bowen and Valenzuela Zapata, 2009; Holloway and Kneafsey, 2004; Trubek and Bowen, 2008).

The re-commodification of food and drink as signifiers of rurality or of regional identity for food tourism has become a common strategy for farm diversification, as part of the development of agri-tourism. Examples can include the opening of farm shops selling produce directly to visitors, pick-your-own fields for fruit and vegetables, on-farm restaurants, farm and vineyard tours, observation rooms for milking parlours, museums and displays illustrating aspects of the production process, farm open days and on-farm accommodation which promises the chance to eat home-grown regional cuisine and to witness food production at close quarters (see, for example, Armesto López and Gómez Martín, 2006; Everett and Aitchison, 2008; Veeck *et al.*, 2006).

More broadly, many rural regions have introduced initiatives aimed at packaging together individual food tourism attractions and promoting the wider district in terms of the gastro-idyll. Armesto López and Gómez Martín (2006) report that in Spain, gastronomic routes have been created offering 'a series of pleasurable moments and activities related to the distinctive elements of a given itinerary: trying products or dishes, guided tours to learn about the agro-industrial production of the resource (either *in situ* or through museum reproductions), purchases in specialist shops and visits to traditional markets' (p. 169). Similar driving, cycling and walking trails linking sites of interest such as cheese producers, vineyards, breweries, cider mills, farm shops, restaurants and pubs have been established in parts of Britain (Storey, 2010) (Plate 4.3). Food festivals are hosted by a growing number of British towns, providing an opportunity to promote local produce, whilst Armesto López and Gómez Martín (2006) note the proliferation in Spain of festivals celebrating particular regional specialities such as mussels, oysters and *empanadas* in Galicia, or tied to traditional events in the rural calendar such as the annual slaughter of pigs in Burgo de Osma or the outdoor barbecues of Catalan spring onions.

Food and drink can also be used to enhance and complement other rural consumption activities. Daugstad (2008) details how brochures for the Norwegian Tourist Association present tourists' engagement with the rural landscape as a multi-sensory experience, including a distinct emphasis on taste, 'with many references to the local food traditions based on local raw materials which may be enjoyed at the [mountain] cabins' (p. 413). One such article focuses on a mountain-top picnic enjoyed by hikers on the island of Alden in western Norway, as Daugstad describes:

> It is presented as an orgy of food, and the view from the 'table' is described in detail, with a shimmering sea, snow-covered mountains on the mainland, and sea-eagles hovering in lazy circles around the island. The orgy itself is presented like this: 'While clouds come and go, happy and food-loving people dish out the treats which they have carried up the mountain to the peak. Cured meat, ham and

Plate 4.3 Food tourism in Devon, England (Photo: author)

> scrambled eggs, home-made pizza, cakes, coffee, soda, strawberries and grapes' ... The event is depicted as a generous feast, not especially of local produce, but one where the food is a major ingredient of the landscape experience.
>
> (Daugstad, 2008: 413–14)

Food tourism is arguably strongest in regions and localities which have managed to retain a distinctive regional cuisine and regional food system, such as Tuscany in Italy, and which are therefore celebrated as the exemplars of the gastro-idyll. In these areas, food tourism may simply build on existing practices, places and events (Miele and Murdoch, 2002). Inevitably, however, the commodification of such events and practices for tourist consumption introduces changes: restaurants get larger and less informal; products are re-packaged to appeal to tourists with rural or regional iconography; and recipes and menus are moderated to cater for tourist palates. In other cases, farmers and producers have switched to new products or production methods that replicate regional traditions, but which had been abandoned in the productivist era (Tregear, 2003). As such, food tourism has contributed to the reinvention of the rural to suit perceptions of the gastro-idyll, replacing a standardized food culture with newly imagined connections between taste and place.

EMBODIED RURAL CONSUMPTION

Bodies in the countryside

The consumption of the countryside is an embodied practice. The preceding section examined how this practice employs the different senses of the body, including sight, sound and taste. This section adopts a more holistic approach, considering how bodies are physically, mentally and emotionally engaged in the consumption of the rural as a multi-sensory experience.

Focusing on the body immediately shifts our perspective. Purely visual consumption of the countryside can be performed from a distance. The drawings that illustrated map covers and guidebooks to seeing the countryside in the early twentieth century commonly adopted what Brace (2003) describes as an 'elevated position', with the viewer looking down on a rural landscape obliquely from above. The same guides encouraged readers to seek out similar viewpoints in their own visual consumption of the countryside (Matless, 1998), and scenic tourism has favoured overlooks and panoramas and viewing towers which place distance between the observer and the scenery. This distance can be more than physical, Brace (2003) commenting that the elevated position allows 'the viewer to be distant from dissenting voices or contested interpretations' of rural landscapes (p. 58). The consumption of the rural through sound or taste can be similarly achieved at a distance, especially if enacted via recordings of nature sounds on CDs and tapes, or via regional specialty foods exported and eaten outside the region.

Fully embodied rural consumption, in contrast, requires the body to be placed in the rural environment; to move through the countryside; to touch and feel the rural landscape. As Crouch (2006) describes, 'the individual thinks and does, moves and engages the world practically and thereby imaginatively, and in relation to material objects, spaces and other people. The individual is surrounded by spaces and does not merely act as onlooker' (p. 359).

The sensuous hunger of the tourist has been fed by more and more innovative commodified rural experiences that offer opportunities to pitch one's body against the rural environment, or to stimulate emotions through getting close to nature. The growing industries of adventure tourism and nature tourism respectively are discussed in more detail later, but first it is worth noting that the multi-sensory embodied character of rural tourism is fundamental to even the most simple of rural leisure activities such as walking.

Walking in the countryside was popularized as a recreational pursuit in the eighteenth and nineteenth centuries, and remains the most widespread activity for visitors to the British countryside (Edensor, 2000). Early advocates celebrated walking as both a romantic and a healthy exercise, but from a relatively early stage writers acknowledged walking as a multi-sensory and embodied experience, as Edensor reveals:

> [The walker's] pores are all open, his circulation is active, his digestion good ... He knows the ground is alive; he feels the pulses of the wind and reads the mute language of things. His sympathies are all aroused; his senses are continually reporting messages to his mind. Wind, frost, rain, heat, cold are something to him. He is not merely a spectator of the panorama of nature, but a participator in it. He experiences the country he passes through, – tastes it, feels it, absorbs it.
>
> (J. Burroughs [1875], 'The Exhilarations of the Road', quoted by Edensor, 2000: 86)

The physicality of walking is felt in each step over different forms of terrain, and in the attrition of effort on body muscles. These physical sensations are intensified in more challenging walks, such as mountain hikes and long-distance footpaths. Edensor (2000), for example, quotes the renowned British walking guidebook writer Alfred Wainwright describing the long-distance Pennine Way in northern England as a 'tough bruising walk, and the compensations are few' (p. 93), and observes that hikers on the 2,500 mile Appalachian trail in the United States, 'not only must overcome physical and mental challenge but also the real dangers of wildlife, disease, hypothermia, assault by humans and severe weather' (*Ibid*). As such, Edensor notes, 'these idealized spartan endeavours are evaluated as producing a superior physical condition and more intense bodily experience to the over-socialized, pampered, slothful bodies of everyday life. The construction of pleasure here relies upon the idea that walking of this type forms character through masculine fulfillment' (*Ibid*).

Walking, however, is not just a physical activity, but is also 'a practice designed to achieve reflexive awareness of the self, and particularly the body and the senses' (Edensor, 2000: 82). The self-reflexivity of walking is vividly and consciously conveyed in Wylie's (2005) narrative account of a walk along part of the South West Coastal Path in England. Wylie's reflexive narrative reveals the walk to comprise, variously, shifting emotions, physical sensations, long periods of solitude and momentary encounters with other people, each of which bring different senses to the fore. At the start of the walk, the path ahead 'resonated not in muscles or bones but in *nerves*' (Wylie, 2005: 235, emphasis in original); later, as the path turns to a thin, muddy strip, Wylie records that,

> Limbs and lungs [are] working hard in a haptic, step-by-step engagement with nature-matter. Landscape becoming foothold. Walkers on the Path very often find themselves in such a close, visual, tactile and sonorous relation with the earth, the ground, mud, stinging vegetation.
>
> (Wylie, 2005: 239)

Further on still, the path emerges on to a cliff-top affording a 'resplendent' coastal panorama, and visual senses kick in: 'It looked somehow too good to be true, as if it had been digitally enhanced and cleaned. It was spectacular. I was all eyes' (Wylie, 2005: 242). Yet, as a corporeal presence in the landscape, Wylie is keen to reject the distance of the rural gaze, suggesting that 'exhilarating encounters with elemental configurations of land, sea and sky are less a distanced looking-at and more a *seeing-with*' (*Ibid*, emphasis in original).

The embodied consumption of the countryside by the casual rambler is hence a process negotiated by the multi-sensory, self-reflexive body as it moves through the landscape. However, as Edensor (2000) notes, this engagement is frequently routinized and structured by embedded cultural norms. Just as scenic tourists reproduce the rural gaze by visiting celebrated viewpoints and taking photographs or buying postcards of views that they have seen on television or in guidebooks, so walkers in the countryside reproduce a geography of bodily movement through rural space:

> The countryside is partly produced by the regular routes which walkers follow ... As a geographically and historically located practical knowledge, walking articulates a relationship between pedestrian and place, a relationship which is a complex imbrication of the material organization and shape of the landscape, its symbolic meaning, and the ongoing sensual perception and experience of moving through space. Thus besides (re)producing distinctive forms of embodied practices (and particular bodies) walking also (re)produces and (re)interprets space and place. Besides inscribing paths and signs in rural space, along with specific patterns of erosion, pedestrian bodies also delineate particular kinds of landscape as suitable for particular kinds of walking.
>
> (Edensor, 2000: 82)

Challenging bodies in adventure tourism

The corporeal sensation of embodied rural consumption is pushed towards the limit in a growing number of rural attractions and activities that entice consumers with the promise of adventure. Collectively labelled as 'adventure tourism', these activities can range from modified traditional pursuits such as rock climbing, canoeing and mountain biking to new technologically enabled experiences such as bungee-jumping, jet-boating and four-wheel-drive safaris. In contrast to more sedate forms of tourism, adventure tourism activities are defined by the emphasis that is placed on the embodied experience, setting out to challenge and stimulate bodies through physical touch and endurance, releasing adrenaline and inducing sensations of excitement, fear, thrill and exhilaration. As Cloke and Perkins (1998) remark, 'adventure tourism is fundamentally about active recreational participation, and it demands new metaphors based more on "being, doing, touching *and* seeing" rather than just "seeing"' (p. 189).

The selection of rural settings for adventure tourism reflects a discursive association of the rural with adventure, drawing on masculinist myths of exploration and the conquest of nature (Cater and Smith, 2003). The rural landscape also provides a dramatic backdrop against which personal heroics of adventure are played out (and captured in photographs and videos), as well as a rugged and tactile environment for testing the body. Cloke and Perkins (1998) hence argue that rural adventure tourism:

> involves exploration of uncharted territory: experiencing the danger and adrenaline rush of past explorers; traveling the untravelable; seeing the unseeable; generally pitting adventurousness, personal bravery, and technological expertise against nature barriers – and winning.
>
> (Cloke and Perkins, 1998: 204)

More prosaically, the fashion for adventure tourism has created economic opportunities for otherwise remote and peripheral rural locations. Places such as Queenstown in New Zealand, Cairns in Australia and Chamonix in France have emerged as prime destinations for adventure tourism, boasting a range of attractions which have commodified rural adventure into 'pay and play' experiences (Cater and Smith, 2003). However, such resorts must also balance the demands of tourist bodies for both thrills and security:

> Adventurous places are typically perceived as marginal locations, frequently rural, that form an alternative to the highly developed west. The reality is that these are as much a part of the global system as the places where participants originate, in that they are well served by air links and have the best hotels and nightlife for '*aprés*-adventure'. What is important is that these places '*look*' like they are at the 'edge of the world', this look adds to that feeling of adventure without compromising the 'safe' regulatory frameworks that holiday makers have come to expect.
>
> (Cater and Smith, 2003: 198)

The combination of risk and security is also evident in the technological facilitation of many adventure experiences. Activities such as bungee-jumping and jet boating were created by technological innovations and quickly commercialized for tourist consumption (Cloke and Perkins, 2002). The technology helps bodies to go further, and to move and act in ways that they could not do otherwise, but it also produces a controlled environment in which adventure is performed, almost as a piece of theatre (*Ibid*).

Similarly, Waitt and Lane (2007) describe the fusing of technology and body in tours of the Kimberley wilderness of Western Australia by four-wheel-drive vehicles. The use of four-wheel-drive vehicles to access the harsh environment of the Kimberley enables more than a version of the motoring tours of scenic tourism, putting the body into the wilderness, away from sealed roads and built-up development. Although the vehicles afford some protection to tourists, movement through the wilderness is still an embodied experience, as Waitt and Lane demonstrate. Not only are bodies exposed to the heat and aridity of the desert, but they also feel the rugged terrain, with one tourist suggesting that driving on unsealed roads helps to forge a connection to place through 'the slower pace that objects flit past the windscreen, the skill of the driver and the roughness of the ride' (Waitt and Lane, 2007: 166).

Tourists interviewed by Waitt and Lane emphasize the thrill of being able to 'go everywhere', yet in practice the routes followed by the four-wheel-drive tourists are highly routinized and structure the consumption of the region and its landscape:

> The routes mapped out in four-wheel drive guidebooks emphasize exploration and adventure. For example, a popular circuit starts from Kununurra along the Gibb River Road to Derby and Broome ... The return journey is along the bitumen of the Great Northern Highway to Kununurra through Fitzroy Crossing, Halls Creek and Turkey Creek. This circuit provides a number of 'side options', including those to El Questro, Kalumburu, Mitchell Plateau, Mount Hart, and the Bungle Bungles. One outcome is an increased flow of four-wheel drive vehicles along the 642 km Gibb River Road from Derby to Wyndham, established in the 1960s as a 'beef road' to enable road-trains from pastoral stations to access these ports. Between 1986 and 1987, the average annual daily flow of vehicles on this road increased from 264 to 329.
>
> (Waitt and Lane, 2007: 161–62)

Thus, Waitt and Lane (2007) argue that the embodied experiences and knowledge of four-wheel-drive tourists act to both refashion and reproduce ideas of the wilderness. The home comforts of the vehicles, and the 'luxury drivescapes' of occasional restaurants, gift shops and hotels, enable tourists to 'drive away' from conventional ideas of the outback, such that 'champagne, croissant and cappuccino are now compatible with experiencing places as wilderness' (p. 167); yet, four-wheel-drive tours can also 'drive home' ideas of outback mythology, as participants interpret their experiences through established discourses and reproduce these through their photographs and stories.

Getting close to nature

The final form of embodied rural consumption involves getting close to nature. Nature-watching has always been part of the visual consumption of the rural, but its distanced perspective is exemplified by ornithologists studying birds through binoculars from discrete hides. Rural tourists are now presented with an array of commodified experiences that offer opportunities to break down the distance with nature, and to get close to both domesticated and wild animals.

These include commodified farm experiences with petting enclosures and children's corners, where visitors are able to touch and feel docile animals such as rabbits and sheep (Daugstad, 2008), as well as working holidays which bring participants into contact with animals through farm work or conservation projects. In another variation, activities such as dolphin swimming offer if not actual touch then at least the sensation of sharing the same space as wildlife. As Besio *et al.* (2008) observe, 'dolphin swimming offers tourists an up-close viewing experience, with the added attraction that viewing takes place by sharing the watery spaces of nature with dolphins ... The embodied experience of dolphin tourism – being in dolphins' spaces not just gazing upon them – produces an intimate connection between the seer and the seen, between humans and nonhuman animals' (p. 1222).

Besio *et al.* (2008) report that the embodied interaction between tourists and dolphins produces an emotional response that contributes to the ways in which tourists understand nature. Yet, once again, these interpretations of nature as encountered are not unguided, but are shaped and informed by the narratives of the tour operators. In particular, Besio *et al.* (2008) argue that a gendered discourse is employed that constructs dolphins both as 'sexy beasts' and as 'devoted mothers', enabling ecotourism operators to 'promote dolphins as creatures that, on the one hand, are wild, sexy, different and exciting, and, on the other hand, are loving, cultural, and the same as humans' (p 1221). Both discourses serve to disguise the act of commodification that enables the tourist experience, but they also contrast with the masculinist discourses that frame adventure tourism. As such, adventure tourism and ecotourism activities such as dolphin swimming reflect two different ways of constructing and encountering the rural as nature, one emphasizing the conquest of wild nature, the other emphasizing a sense of communion with vulnerable nature.

Adventure and encounters with nature are combined in safari tourism, which is an increasingly important economic activity in parts of rural Africa. Safari tourism involves the commodification of African nature through the transportation of primarily European, American or Japanese tourist bodies into the savannah in pursuit of 'an elusive goal, which they themselves would find hard to define – an encounter with nature' (Almagor, 1985: 43). Although Norton (1996) suggests that the experience sought by safari tourists is essentially visual, seeing animals in their 'natural' environment; safaris can produce more embodied encounters, such as an incident described by Almagor in which a group of tourists on foot during a stop on a safari in the Moremi Wildlife Reserve in Botswana are confronted by a wild buffalo. The tourists were potentially in danger, but the risk heightened the experience of a 'direct, spontaneous and first-hand encounter with untamed nature' (Almagor, 1985: 33).

Getting close to nature therefore involves embodied performances both by tourists and by animals. The contingent inter-play is highlighted by Cloke and Perkins (2005) in a discussion of whale and dolphin watching trips from Kaikoura, New Zealand. Here, the sought-after tourist experience relies on the performance of the whales and dolphins, which cannot be guaranteed. The unpredictability of the animals contrasts with the choreography of the tourist performance by the tour operators, including announcements about when sightings should occur, where to view from, and when to take the best photograph. This disjuncture inevitably produces disappointment on the part of tourists when whales and dolphins fail to appear, or when whales fail to deliver the classic vertical flick of the flute.

Crossing human–nature boundaries raises questions about the impact of tourist bodies on nature. Many nature tourism activities are packaged and presented as 'ecotourism', promising minimal environmental impact.

Following the mantra of 'taking only photographs and leaving only footprints' (Waitt and Cook, 2007: 542), ecotourism implies bodily control and self-reflexivity. However, reflecting on their own experiences as ecotourists, Waitt and Cook (2007) recognize that their experiences were always conditional on their bodies, and that 'rather than prefigured or "packaged", our experiences were always open, conditional and creative' (p. 538). Yet, their study of kayakers in Thailand suggests that the power of the ecotourism discourse can substitute self-reflexivity with anxiety about body practice:

> Despite observations that recorded everyone touching the water, plants and rocks, only some respondents acknowledged the material flows between their bodies and the non-human world through touch ... Undoubtedly there are tactile pleasures while paddling from touching and being touched by non-human entities. Yet there is an apparent anxiety about touching.
>
> (Waitt and Cook, 2007: 542)

The discomfort of the ecotourists might reflect a denial of the less pleasant aspects of engagement with nature – 'smells of mangroves and bat urine, being bitten by mosquitoes, squealed at by monkeys, grazed by rocks and burnt by the sun' (*Ibid*: 547) – thus enabling the non-human to continue to be understood in aesthetic terms. It also, though, reproduces the status of touch as a forbidden practice in the ecotourism ethics of care. As such, whilst ecotourism facilitates an embodied encounter with the rural, it can also limit the capacity for the multi-sensory experience to be acknowledged: 'with the almost complete closure of the porosity of tourists' bodies to sensory experiences apart from sight, the myth remains intact of a nature uncontaminated by humans' (Waitt and Cook, 2007: 544).

CONCLUSION

This chapter has explored the consumption of the rural through leisure and tourism. It has shown that these acts of consumption engage multiple senses, of sight, sound, taste and touch. Opportunities for the engagement of each of these senses have been commodified and packaged for consumption by tourists, from pay-to-enter scenic viewpoints, to restaurants serving distinctive local food produce, to petting farms and adventure pursuits demanding physical struggle with the rural terrain. As discussed at the start of the chapter, commodification is a process that increasingly separates the consumed 'signs' from their original referents, and the commodification of rural places and experiences has produced signifiers of a rurality that are eagerly consumed by tourists and visitors, but which have little grounding the actual materiality of the countryside. In this, the commodifcation of rural draws on powerful discourses that have conditioned the sensory experiences of tourists. The rural gaze, for example, reproduces ways of seeing the countryside that guide the observer in where to look, what to see and how to interpret landscape, as well as crucially where not to look and what not to see. The ecotourism myth similarly denies acknowledgement of tactile, embodied encounters between human and non-human, maintaining an imagined separation that supports the idea that tourism consumption can get close to nature without impacting on nature.

In truth, all tourism activity impacts on the rural, because it is through such acts of consumption that the idea of the rural as space of consumption is brought into being, reproduced and modified. The relationship between consumption and the rural is necessarily contingent because the imagined attributes of rurality that are most highly valued by tourists – 'unspoilt' scenery, tranquillity, solitude, pristine nature – are also the qualities that are most vulnerable to the impact of commodification and increased visitor numbers. Excessive commodification can trigger creative destruction, to the detriment of local economies that have been built-up on consumption, but the process is cyclical. As some sites of rural tourism fall from favour, new destinations will be discovered; tired forms of rural consumption will be replaced by new ideas and activities; and the rural as a space of consumption will continue to renew itself.

FURTHER READING

The discussion in this chapter draws on the extensive literature that has been published on rural tourism and recreation. For more on the concepts of commodification and creative destruction see Clare Mitchell's work on St Jacobs, published in two papers in the *Journal of Rural Studies* in 1998 and 2009 (the latter paper with Sarah de Waal). The idea of the rural gaze is introduced by Simone Abram in her chapter in *Country Visions*, edited by Paul Cloke (2003), and Catherine Brace's chapter in the same volume examines how inter-war guidebooks helped to promote particular ways of seeing the countryside. David Matless's article on 'Sonic geography in a nature region', published in *Social and Cultural Geography* (2005) is an excellent discussion of the politics of sound in the countryside, whilst the study of the Tamworth Country Music Festival by Chris Gibson and Deborah Davidson in the *Journal of Rural Studies* (2004) illustrates the aural consumption of the rural through music. Paul Cloke and Harvey Perkins discuss the body in adventure tourism through a case study of Queenstown, New Zealand, in *Environment and Planning D* (1998), whilst Tim Edensor's paper in *Body and Society* (2000) is a good account of embodied practices in walking. The problematic of the body in ecotourism is discussed further by Gordon Waitt and Lauren Cook in *Social and Cultural Geography* (2007).

5

DEVELOPING THE RURAL

INTRODUCTION

The previous two chapters have discussed two different ways of imagining the rural which underpin two different dimensions of the rural economy. The rural as a space of production prioritizes economic activities such as agriculture, forestry, mining and quarrying, in which rural resources are exploited for the production of commodities that are sold on external markets. The rural as a space of consumption, in contrast, is associated with the influx of tourists and visitors seeking multi-sensory experiences of the rural, and contributing to the economy by paying for commodified rural products, activities and places. The discourse of the rural as a space of production was historically stronger, supporting the dominance of primary industries in the rural economy. However, as Chapter 3 demonstrated, the capitalist imperative for profit has led to these industries becoming more and more specialized and integrated, reducing the demand for rural space. As such, large areas of rural territory have in effect become surplus to requirements as spaces of production. In many of these regions, the void has been filled by a new consumption economy based on the commodification of landscapes, customs and experiences for tourism and recreation. Yet, successful commodification requires conformity with the expectations of the rural gaze and other discourses that structure the consumption of the countryside through various embodied practices. Not all rural regions are equally placed to achieve this.

It is this dynamism of the rural economy, and the resulting uneven geography of rural economic performance, that creates the need for 'rural development' policies and strategies. In a pure free market, economic forces would seek out the most profitable arenas of operation, generating cycles of boom and bust for the localities affected. The population would move to follow the economy, which is what happened throughout most of history, as the remains of abandoned farmsteads, villages and mining towns testify. The instability produced by such movements would be catastrophic in the contemporary age. Accordingly, governments instead intervene to try to stimulate and regenerate rural regions that are perceived to be lagging economically, or to be challenged by economic restructuring.

The aims of rural development are therefore relatively simple: sustainable economic growth and improved living conditions, bringing rural areas up to national standards of development, and ensuring that rural regions are attractive places to live and able to contribute positively to the national economy. Strategies for achieving these aims, however, are varied and contested. Approaches to rural development are influenced by political ideology, for example. Social democratic governments will favour direct state action and efforts to improve social equality; whilst neoliberal perspectives favour support for entrepreneurship and market-led solutions (see Chapter 8). Rural development strategies will also differ in response to the particular challenges facing a region, the institutional structures that are in place, and the forms of natural and social capital that can be mobilized and enrolled in actions for development.

All forms of rural development, though, involve a discursive engagement with the rural on at least three levels. First, rural development strategies discursively construct the problems and challenges facing rural regions, which may for instance be presented as geographical peripherality, a lack of competitiveness or poor infrastructure. Second, rural development strategies need to evaluate the capacities that exist in a region – for example unexploited natural resources or a distinctive cultural heritage. Third, a rural development strategy sets out a vision for the future of a rural area, which becomes the objective of its actions.

This chapter examines in more detail the formulation and implementation of rural development strategies, and the ways in which they engage with and reproduce ideas of rurality and constructions of rural space. It focuses on three notable approaches that have been employed to rural development: the modernization paradigm that was dominant in both the global north and the global south during much of the twentieth century; the 'new rural development paradigm' that has supplanted the modernization paradigm in Europe and other parts of the global north, with an emphasis on bottom-up, endogenous development; and the application of participatory rural development in the global south.

MODERNIZATION AND RURAL DEVELOPMENT

The modernization paradigm

Conventional approaches to rural development, in both the global north and the global south, were informed by the concept of modernization. This concept, in turn, was based on the assumption that societies evolve along parallel linear paths from an irrational, technologically limited traditional society to a modern, rational and technologically advanced society (Taylor, 1989). Steps forward along this trajectory resulted from technological innovation, enabling new forms of industrial production and consumption, and supporting the progressive reform of social and political structures and culture. As all societies were considered to be following the same basic path, differences in the economic prosperity of different nations or regions could be explained by them occupying different points along the trajectory of development. As such, poorer countries in the global south were represented as being 'less developed' than richer countries in the global north, with the implication that their development relied on implementing the same policies as had been pursued by advanced industrial nations. Furthermore, as modernity was generally defined in terms of attributes commonly associated with the city – such as industry, rationality and high-culture – urban areas were portrayed as more developed than rural areas, with the development or modernization of rural economies and societies conflated with urbanization.

As such, the modernization paradigm held that the under-development of rural areas, and associated problems of poverty and depopulation, could be addressed by measures aimed at speeding up the spatial diffusion of modernity from cities to the country, and from the global north to the global south. In practice, rural modernization commonly involved four parallel processes. First, *agricultural modernization*, which involved the transition from subsistence farming to commercial agriculture and, subsequently, the mechanization and industrialization of farm processes, the application of agri-chemicals and biotechnology, and the re-organization of the agri-food sector through specialization and integration. As described in Chapter 3, these developments formed the central elements of productivism, and were often underwritten by state support, such as through the Common Agricultural Policy in Europe.

Second, *economic modernization*, which involved the diversification of rural economies away from dependency on traditional industries such as agriculture towards a broader base of 'more modern' industries such as light manufacturing and the high-technology sector (Lapping *et al.*, 1989; North, 1998). Third, *infrastructure modernization*, including electrification and water supply projects, extending telecommunications networks, upgrading road links and the development of regional airports, as well as improvements to the rural housing stock (Matless, 1998; Phillips, 2007; Woods, 2010a). Fourth, *social modernization*, which challenged the superstition and traditional folk cultures of rural societies, perceived as backward-looking, and instead promoted modern rationality and aesthetics, education and social emancipation, and an engaged, responsible and informed practice of good citizenship (Murton, 2007).

In these ways, modernization provided a 'blue-print' for rural development that was applied, with variations, across many parts of the world from the 1920s onwards, reaching a zenith in the post-Second World War period. As discussed further below, the limitations of the modernization paradigm began to become increasingly apparent during the 1970s and 1980s, with a growing critique eventually prompting a turn towards a 'new rural development paradigm' in Europe and other parts of the global north. The modernization paradigm has been similarly criticized in the global south, but remains significant in guiding rural development in several countries, including Brazil, China and India.

Rural modernization in North America and Europe

Rural modernization and state involvement in rural development emerged as conjoined practices in North America towards the end of the nineteenth century. The earliest interventions by the federal government in the United States to support rural development were arguably the establishment of Land Grant Colleges in the 1860s, which as facilities to teach modern agricultural techniques in order to stimulate agricultural development, directly reflected the modernization paradigm. In much of North America at this time, rural development still meant colonization and cultivation by European settlers, and indeed settlement was seen as a modernization strategy: transforming savage wildernesses into civilized societies. Yet, there was also a feeling that the new rural territories represented an opportunity to develop an alternative modernity, free from the vice and corruption of the industrial city. The vision set out by the provincial government in British Columbia, Canada, at the start of the twentieth century, for example, embodied this concept of an alternative rural modernity, as Murton (2007) describes:

> Drawing on traditional agrarian values, the contemporary wilderness ethic, and the discourse of scientific agriculture, they envisioned a countryside inhabited by individual, independent farmers and their families, united by cooperation and community, and living in a spectacular setting of soaring mountains, plentiful game, and easily available recreation. Rural communities would have the latest in modern conveniences, and be linked to the wider world through telephones and good roads. Farms would be run according to the latest research in scientific agriculture, and farm homes would be equipped with modern comforts. The family would be the centre of society, farmers and wives performing their expected roles. This modern countryside would offer an alternative to the version of modernity found in the overcrowded, unhealthy cities.
>
> (Murton, 2007: 60)

At the same time, there was also a recognition that the longer-settled rural regions of the eastern United States were falling behind urban areas in their level of economic development and standard of living. Farms were being squeezed by competition from the Mid-West and California (see Chapter 3), and serious problems of rural poverty and depopulation were starting to emerge. As Lapping *et al.* (1989) summarize, the view of the time was that, 'what was stalling the progress of rural America was its failure to modernize, to take advantage of the efficiencies promised by new technologies, to realize its potential through scientific management and reformed institutions' (p. 26).

The Country Life Movement campaigned for the modernization of rural America, which it presented as involving education, agricultural development, industrialization and new infrastructure. The last objective was boosted during the New Deal in the 1930s, as federal agencies including the Public Works Administration and the Civilian Conservation Corps created employment by constructing bridges, roads, public buildings, hospitals, schools and sewer systems in hundreds of rural communities (Lapping *et al.*, 1989). The most extensive modernization project, however, was to supply electricity to rural areas.

Advocates of electrification regarded electricity as a vital innovation that would stimulate rural development and enable rural areas to catch up with urban areas. As Phillips (2007) notes, they 'believed the government must take an active part in providing electric service to rural areas, for this revolutionary source of power would decentralize industry, restore country life, and "put the farmer on an equality with the townsman"' (p. 27). At the end of the 1920s, only one in ten farms in the United States had electricity. By 1950, after 15 years of work by the Rural Electrical Administration, nine in ten farms had electricity (Lapping *et al.*, 1989).

The need to generate power for rural electricity supplies also formed an impetus for regional development projects. The most famous was the work of the Tennessee Valley Authority, centred on the construction of a series of nine dams along 1,045 km (650 miles) of the Tennessee River, and combining hydro-electric generation, water supply, land improvements, agricultural development and industrialization, especially in the chemical and primary metals industries (Phillips, 2007).

In Europe, rural modernization gained momentum after the Second World War. Alongside the measures for agricultural modernization in the Common Agricultural Policy and national policies, attention was directed at industrialization, infrastructure improvements and settlement rationalization, often in regionally focused programmes. One targeted region was Mid-Wales in Britain, which had lost a quarter of its population between 1871 and 1961. A government inquiry in 1964 presented a picture of a sharply declining farm workforce in an antiquated agricultural sector, limited opportunities for alternative employment, and poor infrastructure. A quarter of houses were without a piped water supply, and 3,000 farms in the region did not have electricity (Woods, 2010a). The Mid-Wales Industrial Association and the Mid-Wales Development Corporation (later amalgamated as the Development Board for Rural

Wales) were established to lead modernization projects including infrastructure improvements and incentives for industrial development. Over 200 new factories were opened between 1977 and 1985, and manufacturing employment increased by 61 per cent, at a time when employment in manufacturing was decreasing nationally in Britain (Woods, 2010a).

The rural modernization project was particularly emphasized by the socialist regimes of Central and Eastern Europe, including the Soviet Union. Socialist policies towards rural areas were informed by Karl Marx's characterization of the 'idiocy of rural life', by which he meant the persistence of superstition and traditional authority in rural society (Ching and Creed, 1997). As such, they aimed both to 'modernize' the rural economy through industrialization and reorganization, and to 'modernize' rural society by replacing authority figures such as priests and landowners with managers and bureaucrats.

The collectivization of agriculture was a key element of rural modernization in socialist states, with individual land-holdings confiscated and amalgamated into large collective farms on which production was mechanized and industrialized. At the same time, factories were constructed in rural small towns to expand non-agricultural employment, and in some states, particularly the Soviet Union, the rural population was forcibly concentrated and resettled in new towns. Pallot (1988), for example, describes the new village of Snov, in present-day Belarus, which was constructed in 1956 to house the 5,000 workers on a collective farm and their dependents, who were previously dispersed among 17 small villages and 800 isolated dwellings. The new village was designed on urban planning principles, with housing provided in blocks of flats and two- or three-storey maisonettes. A central axis through the residential quarters contained public buildings including a school, shops, a social and administrative complex, and an 'architectural monument', and led to a park with a sports stadium and a swimming pool. Residents were provided with a small private plot of land to grow food, but in a break with rural tradition, these were generally located away from the residences on the edge of the village. However, whilst such programmes succeeded in raising the quality of access to some amenities in rural areas, many of the objectives of modernization were not achieved, and the stifling of private enterprise and the dependency on central planning contributed to the entrenchment of widespread rural poverty in Central and Eastern Europe and the former Soviet Union (Shubin, 2007).

Modernization helped to stabilize rural population decline, diversified rural economies and improved accessibility. As such, it created the conditions that enabled counter-urbanization to occur in many countries from the 1970s onwards. However, many in-migrants were attracted by a nostalgic discourse of the rural idyll that positioned the countryside in opposition to modernity. As this paradox became increasingly apparent, so the modernization paradigm was brought increasingly into question. Additionally, the environmental impact of many modernization projects was contested, as was the sustainability of some of the economic benefits delivered. Industrialization had often been achieved by further inward investment from external corporations, making rural areas vulnerable to rationalization and relocation decisions by transnational corporations as globalization progressed (see, for example, Eversole and Martin (2006) on industrial development in rural Victoria, Australia). These criticisms contributed to a turn away from the modernization paradigm in rural development in many parts of the global north during the 1980s and the 1990s.

Rural modernization in the global south

In the global south, the modernization paradigm has framed more than just rural development. The modernization paradigm was used to 'explain' the relative deprivation of countries in Africa, Asia and Latin America, compared with the advanced industrialized economies of Europe, North America and Australasia. The objective of modernization formed the core of national economic policies in countries in the global south, associated with industrialization, urbanization, capital formation and economic liberalization, the promotion of consumerism, and nation-building projects. Within these countries, an internal differentiated geography of modernization was also identified. Mapping of the 'modernization surface' of developing countries by geographers in the 1960s and 1970s found the most advanced levels of modernization in cities, with remote rural areas lagging behind with limited modernization (Potter *et al.*, 2008). The modernization paradigm, though, held that these differences would be reduced as modernization was diffused, justifying the concentration of initial modernization efforts on strategic urban locations.

However, a counter-argument developed during the 1970s which claimed that modernization was failing rural regions of the global south.

Slater's (1974) analysis of Tanzania, for example, showed that rural areas supplied labour to support industrialization and resource exploitation of the core, but derived very little benefit from the activities of the core. The modernization projects that were undertaken in rural regions in developing countries tended to neglect basic needs such as education, health and transport infrastructure, whilst supporting mining and industrial agriculture developments that benefited external investors over local populations. Additionally, modernization facilitated cultural globalization that degraded indigenous cultures. Indeed, in setting Western industrial societies as the benchmark of development for the rest of the world, the modernization paradigm was in practice an agenda for Westernization (Hettne, 1995).

These critiques discredited the modernization paradigm and its underlying assumption of parallel linear progress, labelled by Taylor (1989) as the 'error of developmentalism'. Yet, discourses of modernization have continued to be employed in many countries of the global south, underpinning policies such as industrialization, market liberalization and land reform. In China, modernization has been the driving principle of rural development policy since the start of economic reforms in 1978, advanced through a multi-dimensional approach embracing industrialization, infrastructure development including the construction of new housing, settlement rationalization and urbanization, and the formation of town and village enterprises, as well as social modernization initiatives such as household reform. Since 2006 these objectives have been framed within the over-arching and overtly modernizing policy of 'Building a New Countryside', which includes plans to irrigate ten million hectares of farmland, provide safe drinking water to an additional 100 million rural residents, spend US$12 billion on road construction to connect all rural towns to the road network, to supply an additional ten million rural residents with electricity, and to invest in schools and medical facilities (Long, 2007).

The modernization agenda has significantly transformed rural areas of China over the last 30 years. The rural economy and labour market has been diversified as non-agricultural employment in rural China has increased from around 20 million people in 1978 to over 140 million in 2000, or from 5 per cent of the rural labour force to nearly 25 per cent (Mukherjee and Zhang, 2007). The bulk of non-farm employment is provided in private 'town and village enterprises', which increased in number from around 1.5 million in 1978 to over 23 million in 1993 (Liang *et al.*, 2002), with regional analyses suggesting accelerating growth in the 1990s (Xu and Tan, 2002). Figures suggest that almost half of town and village enterprises are engaged in manufacturing, with significant numbers in the commerce and services and construction sectors (Mukherjee and Zhang, 2007). Industrialization and economic growth has improved living standards, with the per capita income of peasants in Yuhang county in Zhejiang province increasing by nearly 2,500 per cent between 1978 and 1997 (Xu and Tan, 2002). Consumerism has also taken hold. The ownership of colour televisions in Yuhang increased from 1 per 100 rural households in 1985 to 73 per 100 households in 1997; refrigerator ownership increased from zero to 57 per 100 households over the same period (Xu and Tan, 2002).

Yet, disparities in wealth between rural and urban areas have also increased, as have spatial disparities in wealth within the Chinese countryside (Xu and Tan, 2002). These differences mean that whilst the rural industrialization programme was supposed to provide alternative employment for former agricultural workers in rural communities and discourage out-migration, estimated numbers of internal migrants in Chinese cities had increased to around 100 million by the late 1990s (Liang *et al.*, 2002). Furthermore, rural modernization has had a major impact on land use and landscape, including the reconstruction of housing (Long *et al.*, 2009). The accompanying forced resettlement of residents and land reforms have in places met with opposition and protests. These criticisms and challenges have led to a questioning of at least some elements of rural modernization in China (Long *et al.*, 2009).

ENDOGENOUS RURAL DEVELOPMENT

The new rural development paradigm

Writing at the turn of the twenty-first century, van der Ploeg *et al.* (2000) observed that 'rural development is on the agenda precisely because the modernization paradigm has reached its intellectual and practical limits' (p. 395). In Europe and other parts of the global north, the critique of rural modernization has been building since the 1970s. Much of the criticism focused on the problems of over-production, environmental degradation and spatial inequality that suggested that agricultural modernization had come to the end of its usefulness as a development strategy in the countryside as a whole (see also Chapter 3), but questions were also directed at the application of modernization theory in rural development more broadly. As noted above, there was an increasing disjuncture between the modernization agenda and the nostalgia of the rural idyll pursued through counterurbanization and the new rural consumption economy (see Chapter 4); whilst the increasingly foot-loose character of manufacturing in the globalized economy had begun to reverse previous trends of industrialization in many rural regions (Epp and Whitson, 2001; Woods, 2010a).

At the same time, the state-led delivery mechanisms for rural modernization were challenged by the rise of neoliberalism as the dominant political ideology in liberal democracies in the 1980s and 1990s (see Chapter 8). As well as seeking to cut government spending and reduce the involvement of the state in the economy, neoliberalism also questioned the purpose of economic development policy, proposing that the role of the state should not be to lead development, but to foster entrepreneurship and help rural communities to help themselves (Cheshire, 2006).

The result of this critique has been the emergence of a 'new rural development paradigm' (van der Ploeg *et al.*, 2000) that can be defined by three key points of departure from the modernization paradigm (see also Table 5.1). First, there has been a shift in emphasis from *inward investment* to

Table 5.1 Features of the modernization paradigm and the new rural development paradigm

Modernization paradigm	**New rural development paradigm**
Inward investment	Endogenous development
Top-down planning	Bottom-up innovation
Sectoral modernization	Territorially based integrated development
Financial capital	Social capital
Exploitation and control of nature	Sustainable development
Transport infrastructure	Information infrastructure
Production	Consumption
Industrialization	Small-scale niche industries
Social modernization	Valorization of tradition
Convergence	Local embeddedness

endogenous development. The model of spatial diffusion that was central to the modernization paradigm suggested that development of rural regions would come from outside, and rural development policies were often focused on attracting investment from external sources into rural regions. In the new rural development paradigm, the emphasis is on developing the resources found within a rural region, as discussed further below. Second, the mode of delivery for rural development has also shifted from a *top-down approach to a bottom-up model*. Whereas rural modernization was led by the state and involved significant direct state intervention, the new rural development paradigm sees the state as facilitating rural development that is led by rural communities themselves. Third, the structure of rural development policy has moved from *sectoral modernization to territorially based integrated rural development*. The separation of agricultural modernization and rural development as different policy arenas, which prevailed during the post-war period, has been replaced by an integrated approach that combines economic, social and environmental goals within a defined territorial area.

Intrinsic to this paradigm shift is a transition in the discursive framing of rural areas. Rural areas are no longer imagined as lagging regions that require external assistance to move them along a trajectory of development towards a 'modern', industrialized and more urbanized society. Rather, the new rural development paradigm visualizes a differentiated countryside, in which regions have unique social, cultural and environmental resources that can be harnessed in individual and divergent development paths. Rejecting the evolutionary dogma of modernization, it looks back as well as forward:

> Rural development is not just about 'new things' being added to established situations. It is about newly emerging and historically rooted realities that are currently reappearing as rural development experiences *avant la lettre*. Rural development policies should focus on strengthening proven constellations and supporting emergence of new ones. A particularly decisive element will be the combination of the 'old' with the 'new'.
>
> (van der Ploeg *et al*., 2000: 400)

It is this vision that underpins the principles of endogenous development. In particular, Ray (1998) argues that endogenous rural development involves turning 'back' to a territory's indigenous culture and selecting development paths that both employ and sustain the regional culture and environment. He describes this as the articulation of a 'culture economy', or the 'attempt by rural areas to localize economic control – to (re)valorize place through its cultural identity' (Ray, 1998: 3). As such, 'cultures are thus sets of resources available for social and economic control' (Kneafsey *et al.*, 2001: 297), and endogenous development strategies might include actions for food (re-)localization and the promotion of regionally distinctive food and drink products (Fonte, 2008; Kneafsey *et al.*, 2001; Winter, 2003), ecotourism or forms of cultural tourism focused on food or music (see Chapter 4), the commodification of local heritage, the resurrection of traditional craft industries, or the sustainable exploitation of environmental resources (Siebert *et al.*, 2008; see also Box 5.1).

Box 5.1 THE RURAL ECO-ECONOMY

An emphasis on sustainable development is a central characteristic of the new rural development paradigm. At one level this can simply mean rejecting development paths that are environmentally damaging or unsustainable, such as polluting industrial plants, new roads and airports, and chemically intensive agriculture, which all featured in rural modernization. It can also mean seeking to add value to natural resources found in a rural region in a sustainable fashion, and thus to close what Kitchen and Marsden (2009) call the 'eco-economic paradox' – that rural regions 'both hold potentially high ecological value and show persistently low levels of economic activity and welfare' (p. 274).

Kitchen and Marsden (2009) accordingly focus on the development of the 'rural eco-economy'. The eco-economy 'consists of complex networks of webs of new viable businesses and economic activities that utilise the varied and differentiated forms of environmental resources in more sustainable ways. These do not result in a net depletion of resources but provide cumulative benefits that add value to the environment' (p. 275).

Examples of eco-economy initiatives identified by Kitchen and Marsden (2009) in Wales include an organic farm business that produces and sells a single variety apple juice; a community-owned wind turbine; forest mountain bike trails; an equine enterprise staging weekend events for competitive horse rides; and a water sports business offering kayaking, surfing and windsurfing activities that aims to be carbon neutral in its operations.

Further reading: Kitchen and Marsden (2009).

For Ray, the link between territory and culture is essential to the notion of local control and ownership of endogenous development, defining culture as 'a set of *place-specific* forms that can be used to animate and define "development"' (Ray, 1999: 263). Yet, endogenous development also needs to reach out beyond the locality in order to be successful. Ray (1999) accordingly describes endogenous development as 'Janus-faced', needing to look inwards to mobilize local actors including local businesses and community groups as well as local resources, but also looking outwards in order to 'sell' the territory to extra-local consumers and policy-makers. In later work, Ray (2006) suggests that this two-directional model might be more correctly referred to as 'neo-endogenous development', with the prefix 'neo' recognizing the roles played by extra-local actors.

(Neo-)Endogenous rural development therefore depends on the construction and mobilization of networks of actors and resources from both within and outside a rural locality. As the entities enrolled and the relationships fixed will vary between localities, endogenous development gives rise to divergent development paths, some of which will be more successful in achieving their objectives than others. Van der Ploeg and Marsden (2008) portray the process of endogenous development as a web of components including endogeniety, novelty, market governance, new institutional frameworks, sustainability and social capital, which can inter-act and combine in various ways (Figure 5.1). They argue that different regional policies and other structural conditions influence the way in which these components interact, the relative significance that they have in driving development in a territory, and the outcomes that result.

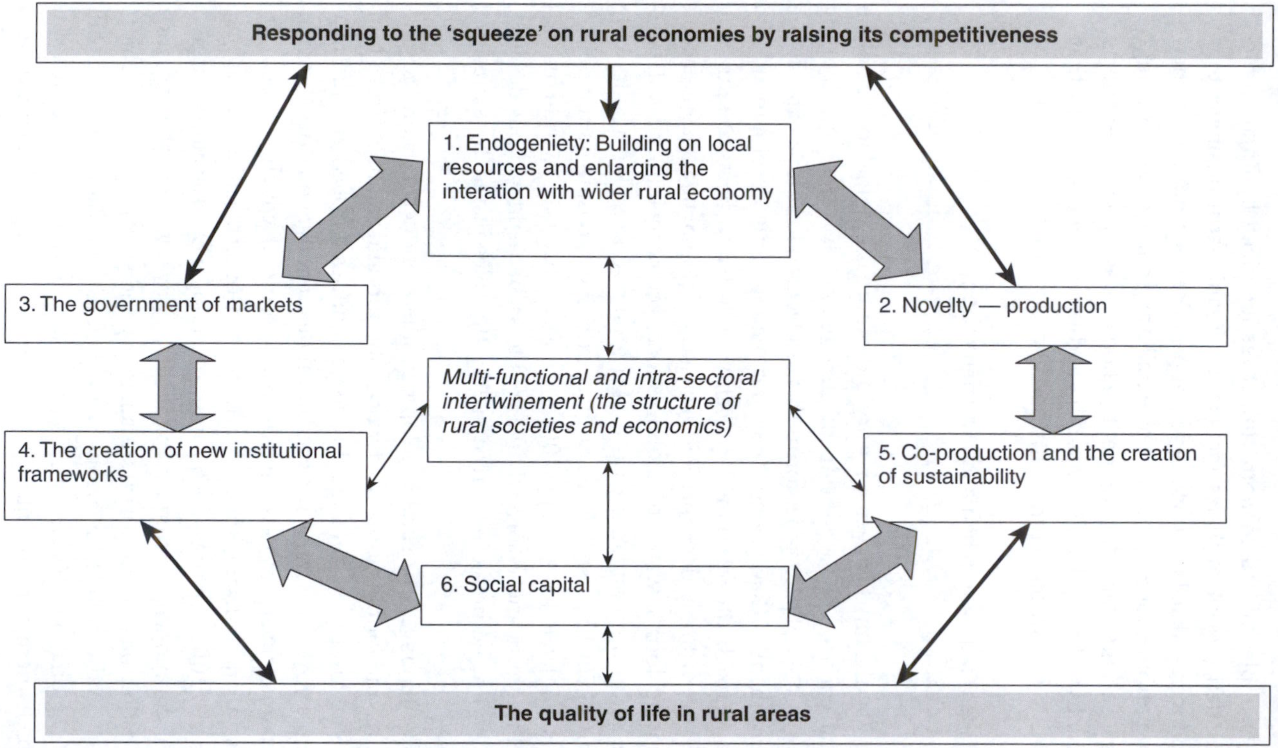

Figure 5.1 The web of rural development (after van der Ploeg and Marsden, 2008) (by permission of Van Gorcum Publishing)

Endogenous development in practice

The implementation of the new rural development paradigm has been most evident in Europe, where it was adopted as the model for European Union rural development policy by the European Conference on Rural Development in Cork, Ireland, in 1996. In particular, the endogenous development approach has been associated with the LEADER programme, which has operated in several incarnations in Europe since 1991. Initially introduced as an experiment to stimulate innovative approaches to rural development at the local level, the size and the coverage of LEADER have increased with subsequent rounds of funding. Similarly, whilst eligibility was originally restricted to rural regions in receipt of regional development support as 'lagging regions' or 'regions undergoing restructuring', since 2006 the LEADER approach has been opened up to all rural areas in the European Union. The model has also been copied in national initiatives, such as PRODER in Spain and POMO in Finland (Moseley, 2003), and the principles of endogenous development more broadly have been replicated elsewhere in EU rural and regional development strategy and in national and regional government programmes.

Critically, LEADER has been implemented through 'local action groups', each covering a territory of fewer than 100,000 inhabitants. These territories were required to have 'some real local identity, rather than simply respect established administrative boundaries' (Moseley, 2003: 12), and as such LEADER has frequently prompted the formation of new institutionalized rural territories, which have in turn become 'brands' for the selling for regions. The local action groups are usually partnerships comprising local government, businesses and community groups, but also work in partnership with communities such that local people are involved in forming and implementing development plans. LEADER groups are also expected to be innovative in their identification of initiatives for rural development and to adopt an integrated approach. As Moseley (2003) explains, 'an example of the latter would be training courses provided for farmers who are keen to diversify, linked to grants to help create off-farm accommodation and linked also to the marketing of the area as a destination for rural tourism' (p. 14).

An illustration of endogenous development in practice can be seen in the case of Pembrokeshire, in south-west Wales, Britain. This was one of the first LEADER areas, with a local action group, 'South Pembrokeshire

Partnership for Action with Rural Communities' (SPARC) set up in 1991 (and succeeded by a second local action group with a larger territorial remit, PLANED, in 2001). SPARC set up a network of 'village-based community associations' which 'were central to SPARC's underlying aim of giving local people the chance to develop their own communities, economically, socially, environmentally and culturally' (Moseley, 2003: 45). Each association carried out a 'community appraisal' to consult residents and establish local needs, priorities and capacities, and subsequently involved residents in drawing up local action plans. These were integrated with a number of strategic plans developed by SPARC across the territory. The result has been a series of initiatives for rural development implemented over nearly two decades, aimed at attracting tourists and providing opportunities for local businesses. Key examples include cultural tourism initiatives such as restoring historical sites, way-marking footpaths and producing interpretative panels; a local products initiative, supporting the production of local food products by linking producers and purchasers, providing training and promotional activities; a 'Quality in Business' scheme offering training, advice and facilities for small local enterprises; and a 'Demonstration Farm Review and Development Scheme', which 'encouraged the development of whole-farm business plans linking training, diversification and conservation audits to funding for business and environmental improvements' (Moseley, 2003: 16).

The principles of endogenous development have also been introduced into rural development in Australia, Canada, New Zealand and the United States, although without the coherence of an overarching framework as offered by EU policy and the LEADER programme (Brennan *et al.*, 2009; Cheshire, 2006; Bruce and Whitla, 1993). As such, endogenous development in these countries depends even more heavily on local initiative, innovation and entrepreneurship. This is demonstrated by the cases of two pseudonymous rural districts in Queensland, Australia (Cheshire, 2006; Herbert-Cheshire, 2003). In the first, Woomeroo, an action group was established to fight plans to downgrade local rail and court services, but having failed in these objectives changed tack and 'rather than reacting to the closure of local services with petitions and lobbying tactics, the group channelled its energies towards a more proactive type of response, which involved working in accordance with federal and state government policies to access funding for community and economic development activities'

(Herbert-Cheshire, 2003: 463). A workshop attended by around 80 residents identified eight projects that were taken forward, including the restoration of the town's historic courthouse as an information centre and the construction of a pool complex to encourage tourism, and investment in a machine to make cement blocks using sand dredged from a local weir (Cheshire, 2006).

In the second district, Warmington, a revival committee comprising seven elected representatives from local government and agriculture was formed in response to pressure on traditional farming sectors. It focused on the development of a linseed/flax industry in the area as its major project, eventually supported by a government grant. However, Cheshire (2006) reveals that bidding for funding involved compromises that created dissent among local actors in Warmington, and as such her case studies demonstrate that rural development is always a political process.

Critiques of endogenous development

Endogenous rural development has become the prevailing approach in most of the global north, and has in general delivered more sustainable forms of development to rural regions than that offered by the modernization paradigm. However, there are three key critiques of endogenous development that have been advanced by various commentators. First, the capacity of endogenous development to tackle issues of fundamental structural disadvantage has been questioned. Analyses of relative economic performance in England (Agarwal *et al.*, 2009) and the United States (Isserman *et al.*, 2009) have shown that geographical position, transport infrastructure, economic structure and the education levels of the population are still important factors in determining relative prosperity. These inequalities are not easily addressed through bottom-up territorial initiatives, and as Markey *et al.* (2008) argue in the case of northern British Columbia, there is still a case for some state-led infrastructure projects to create appropriate conditions for endogenous development.

Second, the capacity of local communities to successfully engage in endogenous development is also uneven. Numerous studies have highlighted the importance of social capital in bottom-up initiatives (Árnason *et al.*, 2009; Magnani and Struffi, 2009), yet levels of both 'bonding social capital' (coherence within communities) and 'bringing social capital' (the ability to form networks outside the community) vary significantly (Árnason *et al.*, 2009; Brennan *et al.*, 2009; Woods *et al.*, 2007) (see Box 5.2). This means that the places best equipped for endogenous development are settled rural communities with professional middle-class residents, which are also arguably among the areas least in need of development.

Third, endogenous development may not be as inclusive as its advocates claim. Although the bottom-up approach is often presented as empowering communities, actual levels of public participation can be

Box 5.2 SOCIAL CAPITAL

Social capital refers to the collective resource that is created by the formal and informal ties and interactions between members of a community. Along with other types of 'capital', including financial capital and human capital (people), it contributes to a community's collective capacity to act. Although he was not the first to use the term, social capital is now commonly associated with the American political scientist Robert Putnam. In a study of Italy, Putnam (1993) proposed that the strength of social capital based on centuries of civic tradition explained the relative political stability and economic prosperity of the north of Italy compared with the south. His follow-up work, *Bowling Alone*, argued that American society had been damaged by the erosion of social capital and the rise of individualism, and that action was required to restore social capital (Putnam, 2000).

Putnam (2000) describes two types of social capital: *bridging* capital and *bonding* capital. Bridging capital refers to social networks that link members of a community with other communities, or other scales; whilst bonding capital refers to the strength of social connections within a community. Bonding social capital hence assists in building community solidarity and helping members of a community to act together, whilst bridging social capital enables a community to enrol the support of external actors and to put its case at a higher scale. Bridging and bonding social capital can also reinforce each other, such that when bridging and bonding social capital are both high, effective community action or entrepreneurialism results; whilst low bonding capital and low bridging capital is reflected in individualism and apathy (Flora *et al.*, 2008).

Social capital has become a widely used concept in both policy and academic discourses of rural development (see for example Árnason *et al.*, 2009; Brennan *et al.*, 2009; Flora *et al.*, 2008; Magnani and Struffi, 2009). These accounts tend to emphasize two assertions. First, that social capital is an important tool in endogenous rural development, enabling communities to help themselves; and second, that in order to address inequalities in rural development, attention should be paid to building social capital in rural communities where it is currently weak.

However, there is also an extensive critique of social capital as an idea. Critics have argued that Putnam's thesis is normative rather than analytical, in that it ignores evidence that does not fit, as well as accusing social capital theory of downplaying the significance of wider social and economic structures and of failing to demonstrate causality (Anderson and Bell, 2003). There is also a 'dark side' of social capital, with high levels of social capital measured according to Putnam's method associated with conformity, paternalism, inequality and the marginalization of alternative lifestyles (Schulman and Anderson, 1999; Anderson and Bell, 2003) (see also Chapter 6).

Further reading: Anderson and Bell (2003), Flora *et al.* (2008), Putnam (2000).

low, or can tail off rapidly. As such, endogenous development can in some places be accused of concentrating influence with local elites, or with rural development professionals (Cheshire, 2006; Kovacs and Kucerova, 2006; Shucksmith, 2000; Woods *et al.*, 2007). Moreover, Shucksmith (2000) argues that the 'territorial approach tends to mask inequalities and power relations between social actors within a "community" by employing a consensus perspective' (p. 209). He suggests that by defining communities as territories, rather than as groups of individuals, the endogenous development discourse obscures differences of class, ethnicity, gender and age. Consequently, 'endogenous development only has the potential to challenge processes of exclusion if it empowers those without power (and this is not necessarily the same as empowering the local against the external)' (*Ibid*: 210) (see also Árnason *et al.*, 2009; Shortall, 2008).

REORIENTING RURAL DEVELOPMENT IN THE GLOBAL SOUTH

From colonial development to post-colonial development

The transformation of approaches to rural development in the global south has been no less significant than that experienced in the north, but considerably more dramatic, dynamic and contested. There are broad similarities in the trajectories followed in both parts of the world, but the experience of the global south has been shaped by the particular circumstance of countries in Africa, Asia and Latin America: the greater challenge of development in regions without basic infrastructure, with endemic poverty and high levels of dependency on agriculture, and poor levels of literacy and standards of health; unstable and shifting political-economic regimes and ideologies; and the colonial legacy, both material and discursive.

Rural regions in the global south were first imagined by colonial powers as spaces of exploitation and extraction. Development took the form of mining ventures and plantations of cash crops, and the infrastructure required to support these activities, and was for colonial economic gain. There was no concern for the development of the region itself, with indigenous populations being paid little regard, except to be conscripted for labour, or displaced or eradicated as obstacles to development. The legacy of this discourse is vividly portrayed in Eduardo Galeano's polemical history of Latin America:

> Latin America is the region of open veins. Everything, from the discovery until our times, has always been transmuted into European – or later United States – capital, and as such has accumulated in distant centers of power. Everything: the soil, its fruits and its mineral-rich depths, the people and their capacity to work and to consume, natural resources and human resources ... To each area has been assigned a function, always for the benefit of the foreign metropolis

> of the moment, and the endless chain of dependency has been endlessly extended. The chain has more than two links. In Latin America it also includes the oppression of small countries by their larger neighbors and, within each country's frontiers, the exploitation by big cities and ports of their internal sources of food and labor.
>
> (Galeano, 2009: 2)

Even post-independence, the rural regions of the global south continued to be discursively positioned as spaces of resources exploitation for transnational corporations engaged in mining, oil exploration or the agri-food sector. The activities of agri-business in the global south has been touted as a model of rural development, creating markets for local produce and investing in modernization, but agri-business has also been criticized for undermining local peasant economies and some of its actions, particularly with respect to the promotion of hybrid seeds and GM crops, have been attacked as neo-imperialist (Kneen, 2002; van der Ploeg, 2008).

Initially, at least, post-independence governments often did little to change this situation. As discussed above, the modernization paradigm held that development would trickle-out to rural areas, thus justifying the absence of an overt rural development strategy. Nationalist leaders emphasized the discursive integrity of rural regions to the new state territories, but nation-building projects such as dams, new roads and land reform were top-down policies that commonly disrupted rural systems and brought limited direct benefit to rural communities, tending instead to reinforce the perceived urban-bias in economic development (Potter *et al.*, 2008) (see Box 5.3).

It was only in the 1970s that substantial programmes for rural development appeared, framed by a new discourse of 'integrated rural development' that was promoted by bodies such as the World Bank (Potter *et al.*, 2008). The integrated rural development discourse recognized rural regions as complex systems in which issues of social and economic development were inter-connected. It promoted schemes that tended to focus on improving agricultural productivity, and also included actions to improve rural health care, education and transport, as a multi-faceted approach to poverty alleviation. These objectives were pursued through state intervention and investment, such that integrated rural development formed 'an attempt to mobilize public sector resources in an integrated

Box 5.3 RURAL DEVELOPMENT IN INDONESIA

The trajectory of rural development strategy in Indonesia illustrates the different approaches that have been attempted in many countries of the global south, and the critical importance of the state and political ideology in shaping the principles and practice of rural development. During the Dutch colonial era, an extensive plantation economy was developed in Indonesia producing traditional crops such as sugar, coffee, tobacco, tea and cinchona for export, primarily to Europe, with rapid growth between 1870 and 1914. New plantation crops including rubber and oil palms were also introduced in the early twentieth century. Production of the traditional plantation crops peaked in the 1910s, as exports were hit by the inter-war global depression.

Following independence in 1945, the new nationalist government promoted land reform in line with its ideology of *pancasila*, which 'declared that land has a social function and must therefore be controlled by the state' (Kawagoe, 2004: 185). In practice, however, land distribution was limited, the structure of the Indonesian rural economy continued much as during the colonial era, and the primary consequence was stagnation.

From 1966 onwards, shifts in government ideology produced a greater openness to economic liberalization. Between the mid-1960s and 1980s, Indonesian agriculture experienced considerable growth, especially in food crops including the traditional staple crop, rice. A significant reduction in rural poverty was also recorded during this period. Rural development strategy was led by US-trained technocrats and included the rehabilitation of irrigation facilities, enhanced agricultural extension (training and advice) services, and new credit facilities for farm investments. Subsidies were provided for imported fertilizers and pesticides, which were distributed to farmers through village organizations.

Yet, Kawagoe (2004) argues that through this era, 'rural development policies in Indonesia were characterised by extensive state intervention, neglect of the traditional sector, and mistrust of markets' (p. 199). While production of food crops, especially rice, was supported, traditional export crops such as tea were neglected due to anti-colonial sentiment in the government. Indeed, Kawagoe argues that with the exception of rice, agricultural development was marginalized by an emphasis on industrialization in government policy.

One side effect of this policy was the increasing significance of smallholders in producing crops such as palm oil, coffee and cocoa, but the lack of aggregation has meant that small producers have found it difficult to trade in the world market on favourable terms. Limited capital to invest in their farms has also constrained small producers, but this has eased with the growth of community-based microfinance initiatives since the 1980s, often set up by NGOs. These include the 'Bank Desa', which give loans to villagers without taking collateral (Shigetomi, 2004).

Further reading: Kawagoe (2004), Shigetomi (2004).

manner to try to channel resources to rural development and poverty reduction' (Zezza et al., 2009: 1299). However, the state-led approach was also a weakness:

> [Integrated rural development] projects tended to be centrally designed and developed through a top-down approach, which often failed to consider local conditions, develop local capacity and foster local participation. As such, the projects were generally not sufficiently flexible to allow for differences across region or households in livelihood strategies.
>
> (Zezza *et al.*, 2009: 1299)

The lack of local participation meant that integrated rural development was frequently reliant on central state direction and funding, which became increasingly squeezed in the 'debt crisis' of the 1980s and 1990s as structural adjustments were imposed on debtor countries by the World Bank and the International Monetary Fund (IMF). These in turn became the mechanisms for introducing neoliberal reforms, which sought to reduce the role of the state. The rise of neoliberalism in the global north had similarly made governments and international aid donors sceptical of state intervention, and more concerned about efficacy and accountability (Potter *et al.*, 2008). Consequently, non-governmental organizations (NGOs) became the preferred vehicle for delivering rural development, including international aid agencies such as Oxfam. NGOs operated within rural communities and engaged with local civil society, thus enacting a new discursive turn in rural development in which rural regions were recognized as composed of communities, calling for rural development to be locally grounded and territorially differentiated.

Community-centred rural development

The turn towards community-centred rural development in the global south has been presented by Shepherd (1998) as a 'new paradigm' in rural development, mirroring many of the characteristics identified in 'new rural development paradigm' in the global north. It emphasizes a 'bottom-up' community-led approach, the valorization of local endogenous resources, and a holistic perspective on social, economic and environmental objectives. As indicated in the discussion above, the approach emerged from a critique of the integrated rural development model, a critical moment being the conversion of the World Bank in the early 1990s, persuaded by successful examples of community engagement in natural resource management and in the administration of social funds (Binswanger, 2007).

Equally, however, the new paradigm has intellectual foundations, especially in the work of Robert Chambers (1983, 1993), who outlined its theoretical principles. In particular, Chambers argued that rural development should identify and emphasize the priorities of small farmers, rather than the priorities of development practitioners and researchers – in other words, it should 'put farmers first'. He formulated the concept of 'participatory rural appraisal' (PRA) as a mechanism for establishing community priorities (see Box 5.4), identifying starting points for rural development that 'are at once dispersed, diverse and complicating' (Chambers, 1993: 120). The approach hence emphasizes providing 'baskets of choices' for rural development strategies rather than blue-prints, and seeks to combine local traditions and Western science (Potter *et al.*, 2008).

Box 5.4 PARTICIPATORY RURAL APPRAISAL

Participatory rural appraisal (PRA) is a methodology developed by Robert Chambers, professor in the Institute of Development Studies at the University of Sussex, and a major influence on contemporary rural development theory in the global south. PRA attempts to put into practice the principle proposed by Chambers (1994) of 'putting farmers first' in rural development. As such it aims to help members of rural communities to assess the realities of their lives and conditions, to formulate plans for action, and to implement and monitor initiatives.

PRA does not prescribe any one method, but offers a range of techniques to facilitate information-sharing, discussion and analysis, including visualization tools. As Korf and Oughton (2006) describe, 'PRA is often conducted in workshops where large parts of a rural community meet in public forums or in smaller groups to discuss and exchange ideas under the facilitation of external moderators. Information is shared between insiders (the villagers) and outsiders (the planners)' (p. 284).

Korf and Oughton also note critiques of PRA, especially the assumptions of social learning and non-coerced communication (Kapoor, 2002; Leeuwis, 2000). They acknowledge that 'there seems to be the assumption that lack of knowledge would impede local development and that, if local knowledge could be properly made use of, this would lead to more locally adapted solutions' (Korf and Oughton, 2006: 284), and note that the unproblematic treatment of 'local knowledge' in this assumption contrasts with the actual dynamics of PRA events. As such, they propose that PRA should be seen as a bargaining process, where different voices are heard and outcomes negotiated.

Further reading: Chambers (1994), Korf and Oughton (2006).

The formulation of rural development strategies in the new paradigm is additionally informed by the sustainable livelihoods framework (Neefjes, 2000; Scoones, 1998). This draws on Chambers and Conway's (1992) conceptualization of a livelihood as 'the capabilities, assets (stores, resources, claims and access) and activities required for a means of living' (p. 7), and their proposal that a sustainable livelihood is one 'which can cope with and recover from stress and shocks, maintain or enhance its capabilities and assets, and provide sustainable livelihood opportunities for the next generation; and which contributes net benefits to other livelihoods at the local and global levels in the long and short term' (pp. 7–8). The sustainable livelihood framework offers an analytical tool for understanding the interaction of different factors that affects the components of human, natural, physical, social and financial capital that comprise the livelihood assets of an individual and their ability to achieve a sustainable livelihood (Neefjes, 2000; Potter *et al.*, 2008). In so doing, it can inform decisions about rural development interventions aimed at improving individuals' living standards, and thus also assists the paradigm in getting around the reification of community that has been identified in endogenous development in the global north.

One key challenge for the new paradigm compared with the implementation of endogenous development in the global north is the relative weakness of local scale institutions in many parts of the global south, labelled as the 'institutional vacuum' (Zezza *et al.*, 2009). The consolidation of a institutional infrastructure has accordingly become an objective of rural development strategies in itself. This includes not only developing local government structures, but also establishing civil society village organizations and producer associations. Bernard *et al.* (2008), for example, record that the presence of village organizations in Burkina Faso increased from 22 per cent of villages in 1982 to 91 per cent in 2002, and that in Senegal, the proportion of villages with at least one village organization rose from 10 per cent to 65 per cent over the same period. These village organizations are diverse in form and focus, including both market-orientated organizations involved in activities such as processing and marketing, livestock breeding and animal husbandry, horticulture, cotton production and handicrafts; and community-oriented organizations with activities such as the provision of credit, managing collective cereal banks, environmental management, water management and sports and social activities.

The evidence for the contribution of local institutions to material gains in rural development, however, is more mixed. D'Haese *et al.* (2005) describe how a local woolgrowers' association in Transkei, South Africa, assisted farmers in lowering costs through collective action and increasing prices. Yet, Bernard *et al.* (2008) note that some local institutions have been criticized for corruption and for excluding poorer community members. With reference to Burkina Faso and Senegal, they report that village organizations have become a major channel for governments and aid agencies in reaching the rural poor, but conclude that 'institutional richness does not (yet) translate into substantial material benefits that could make a difference in rural development and support the competitiveness of a smallholder sector' (p. 2202).

By definition, rural development initiatives within the new paradigm can address a range of foci, responding to local priorities. Primarily, though, the approach supports the use and valorization of local resources, thus shadowing the model of endogenous development in the global north. Many actions have concerned agricultural improvements, corresponding with the economic centrality of agriculture and needs to improve nutrition and food security, as well as opportunities to access external markets. Other actions can concentrate on local cultural and environmental resources. For example, Jackiewicz (2006) records initiatives by community associations in the village of Quebrada Grande in Costa Rica, including an ecotourism venture based on preservation of the green macaw and a fish-farm cooperative operated solely by women.

The sustainable commodification of the natural environment in such initiatives is particularly significant as it represents a further discursive shift, presenting nature in the global south as an asset to be managed and preserved rather than as a resource to be extracted and exploited. This reflects growing awareness that global sustainability goals require the protection of environmental resources in the global south, and that this is best achieved if communities can be persuaded that there is greater economic gain in preserving the environment and wildlife than in extractive activities. A number of initiatives have been established to support this approach, generically known as payments for environmental services (Engel *et al.*, 2008). One example is the CAMPFIRE programme in Zimbabwe, through which rural authorities sold the rights to access wildlife to entrepreneurs for safari hunting and ecotourism. It was designed 'to stimulate the long-term development, management and sustainable use of natural resources in Zimbabwe's communal farming belt' (Frost and Bond, 2008: 777), and gave rural communities custody over wildlife resources and the right to benefit from their use. Through the programme, safaris and ecotourism became a major source of income for communities in the area, and created an incentive for continuing sustainable management of the resource.

Community-centred rural development approaches have had a significant impact in shifting the discourse and practice of rural development in the global south, yet they are still not universally applied, nor are they free from problems. As with endogenous rural development in the global north, the long-term effectiveness of the paradigm is yet to be proven. Critical to any success is the appropriate and accurate representation of rural communities, their interests and their workings, and thus the discursive framing of rural development continues to be important. Interestingly, Bebbington (1999) suggests that one reason for the failure of rural development projects in marginal areas of Latin America is that 'they simply misperceive the way people get by and get things done' (p. 2021). In particular, Bebbington argues that 'a large part of the problem is that interventions work with ways of seeing the world that continue to crunch rural livelihoods into the category of agricultural and natural resource-based strategies' (*Ibid*). Such problems may be overcome by models such as the sustainable livelihoods framework, but they illustrate the power of discourses of rurality within rural development, and the danger of discursive misrepresentation for rural development strategies.

CONCLUSION

Approaches to rural development have evolved over the course of the last century, shifting with political ideologies, economic conditions, the trial and error experimentation of development professionals and institutional learning. Changes in rural development strategies have also reflected transformations in the discursive framing of rurality, and hence in the ways in which the challenges and capacities of rural areas are understood, and visions for their future constructed. The modernization paradigm, which conceptualized rural areas as lagging behind urban areas in their level of progression towards a 'modern', technologically advanced and industrialized economy and society, has given way to new paradigms in both the global north and the global south that emphasize bottom-up endogenous development, involving local people and utilizing local resources (although ideas of modernization are still influential in rural development in many areas, especially in industrializing countries such as China and India).

On the one hand, this transition has brought to the fore the differentiation of rural regions. The new rural development paradigms in the global north and the global south both recognize that rural regions are composed of many different communities, each with their own identities, needs, aspirations and capacities to act. Rural development strategies are no longer about imposing solutions or blue-prints for development, but are about helping rural communities to identify their own objectives and implement their own plans for development. This transition in itself reflects a significant change in the discursive framing of rural areas, from being perceived as 'backward' regions in need of assistance, to being represented as capable and self-reliant regions with the capacity to develop themselves.

At the same time, the new approaches in rural development also stress the inter-connection of rural regions. This in part acknowledges the deepening of globalization and the integration of rural localities around the world into global networks; but, at a more discursive level, it also represents the idea that different rural areas can share similar experiences, and therefore are able to inspire each other. The concept of mutual learning between localities has been a central element in the European Union's LEADER programme, for example, acting as a stimulant to endogenous development (High and Nemes, 2007; Ray, 2001). However, there is also growing recognition of the potential for the transfer of ideas and examples between rural communities in the global north and the global south, and vice versa.

The spheres of rural development in the global north and the global south have commonly been held apart, largely because of the apparent differences in the problems that are faced. Whereas in the north the challenge for rural development is to help rural areas adjust to the declining significance of agriculture and other once staple industries, the challenge for rural development in the south is more stark: in many developing countries at least a third of the rural populations lives below nationally defined poverty lines (rising in some countries to more than half the rural population) (Potter *et al.*, 2008); there are widespread problems of chronic undernourishment, disease and lack of access to clean water supplies; and agriculture, including subsistence farming, continues to dominate the economy of most rural regions.

Yet, in spite of these differences, there are also commonalities and opportunities for the exchange of ideas and approaches between the global north and the global south, as can be illustrated by three short examples. First, payments for environmental services as part of a strategy for sustainable development have been utilized in both the north and the south, with models such as ecotourism being adopted in both hemispheres (Engel *et al.*, 2008). Second, as community-centred approaches to rural development take root in the global south, there is growing interest in the model of territorially focused development represented by the EU's LEADER programme, especially in Latin America (Zezza *et al.*, 2009). Third, as rural development practitioners in the global north struggle with the challenge of overcoming the exclusion of marginalized groups within communities, writers such as Korf and Oughten (2006) have suggested there is scope for the application of ideas such as participatory rural appraisal and the sustainable livelihoods model in European rural development.

As such, rural development should be understood as a dynamic process that is always contingent on the coming together of different local and extra-local actors and elements. It is therefore an inherently political process, involving constant negotiation and contestation, with the discursive framing of rurality, the representation of community, and the envisioning of rural futures, lying at the heart of the debate.

FURTHER READING

Most literature on rural development focuses either on the global south or on the global north, and there are very few papers and books that engage with both. Benedikt Korf and Elizabeth Oughton's paper in the *Journal of Rural Studies* (2006), asking what Europe can learn from the south, is one of the few to cross the line. *Geographies of Development* (2008) by Robert Potter, Tony Binns, Jennifer Elliott and David Smith provides a good introductory overview of trends and challenges in rural development in the global south, whilst the paper by Alberto Zezza and colleagues in *World Development* (2009) is an excellent up-to-date discussion of some of the key issues and debates in rural development in the south. As noted in this chapter, the recent paradigm shift in rural development in the global south has been strongly influenced by the work of Robert Chambers, particularly the books *Rural Development: Putting the Last First* (1983) and *Challenging the Professions: Frontiers for Rural Development* (1993), which are good starting points. For the global north, the paper by Jan Douwe van der Ploeg and colleagues in *Sociologia Ruralis* (2000) provides a general introduction to the idea of the new rural paradigm. Two further papers in *Sociologia Ruralis* provide accessible examples of endogenous development in practice, in both cases drawing on evidence from Wales: Moya Kneafsey, Brian Ilbery and Tim Jenkins (2001) explore the valorization of the 'culture economy' in west Wales, whilst Lawrence Kitchen and Terry Marsden (2009) discuss the development of the rural eco-economy.

6

LIVING IN THE RURAL

INTRODUCTION

In Halfacree's three-fold model of rural space, introduced in Chapter 1, the third dimension is formed by the everyday lives of the rural, which interact with formal representations of rurality and the spatial practices of rural localities. Everyday life in the rural may be shaped by the socio-economic structures of rural localities and informed by representations of rurality, but as Halfacree notes, these aspects 'never completely overwhelm the experiences of everyday life' (2006: 51–52). People living in rural areas make the rural through their own routine practices and performances (discussed further in Chapter 7), through their lifestyle choices, and through their interactions with other rural residents, both human and non-human.

Experiences of living in the rural vary enormously between individuals. The lifestyle promised by the discourse of the rural idyll (see Chapter 2) may be enjoyed by a lucky few, but for the vast majority of rural residents around the world, life continues to be a struggle with everyday issues such as the demands of work and family life, or problems with money, health, crime, loneliness or alienation. In many cases, the lifestyles of rural residents may seem little different to those of urban residents. In other cases, a rural setting creates its own distinctive problems of isolation, lack of employment opportunities, pressure to conform, or difficulties in accessing vital services. There are large differences between the meaning of rural life as experienced in communities in the global south, and its meaning in more prosperous rural regions of the global north. Yet, even within a single rural community, rural life can be interpreted and experienced in many different ways: cherished by some, detested by others.

This chapter examines the ways in which rural life is both conceptualized and experienced, and investigates how notions of living in the rural have been disrupted and remade by processes of restructuring. The first section discusses the importance of ideas of community and belonging to place in the discursive construction of rural life, and looks at how these concepts are given material expression in the lives of rural residents. It notes that the concept of the 'rural community' has traditionally been associated with stability, coherence and security, but also recognizes that the concept has a dark side, and that close-knit rural communities can reproduce unequal power relations and enforce conformity, excluding those who are deemed not to fit or who exhibit supposed 'deviant' characteristics or behaviour.

The second section of the chapter considers some of the ways in which rural communities, as traditionally imagined, have been destabilized by the effects of social and economic restructuring, including both out-migration and in-migration. In particular, it focuses on the impact of counterurbanization in Europe and North America, examining the dynamics though which in-migrants become integrated into rural communities and develop attachment to place, but also probing the potential for tensions to arise between long-term residents and new arrivals over aspects of community identity and rural life. The section then proceeds to investigate the rise of mobile and transitory rural communities, whose attachment to fixed rural places is limited, including second home owners and migrant workers. Additionally, it is argued that the increased mobility of rural populations has stretched the spatial expression of rural communities, producing new configurations that can be trans-local and even transnational, for example, through rural migrant workers in towns and cities or foreign countries who maintain identity with and participation in their home rural community.

The final section of the chapter recognizes that people are not the only inhabitants of rural areas, but that rural space is shared with plant and animal communities and that coexistence with nature is part of the discursive construction of rural life. It discusses how rural communities have learnt to live with nature, including large carnivores, and how non-human life produces its own geographies of being within the countryside.

COMMUNITY, BELONGING AND RURAL PLACE

Conceptualizing rural community

The notion of community has long been synonymous with rural life. As rural sociologists endeavoured during the early twentieth century to distil the essence of rural society, as distinct from urban society (in much the same way that rural geographers sought to define the characteristics of rural space (see Chapter 2)), they repeatedly returned to theorizations that associated rurality with forms of social interaction based on a stable and structured community. This work was strongly influenced by the contrast drawn by the German sociologist Ferdinand Tönnies between *gemeinschaft* and *gesellschaft* as two alternative forms of social organization. In Tönnies's model, originally published in 1887, *gemeinschaft* refers to community as a social grouping based on mutual bonds, a feeling of togetherness and collective goals, whilst *gesellschaft*, or society, is described as being based on individualism, with collective identity and action only sustained as long as they serve an instrumental purpose in supporting individual objectives (Tönnies, 1963).

Tönnies identified *gesellschaft* with modern urban social formations, and thus, by implication, assigned *gemeinschaft* as characteristic of more traditional rural settings. However, Tönnies himself did not suggest that *gemeinschaft* and *gesellschaft* were defining features of rural and urban social formations in themselves, and indeed discussed the family unit and pre-modern neighbourhoods (which could be urban as well as rural), as exemplars of *gemeinschaft*. It was later rural sociologists and rural geographers who drew on Tönnies to propose a 'taxonomy of settlement patterns' (Newby, 1977: 95) in which rural settlements were defined by the practice of *gemeinschaft* or community (Panelli, 2006). As Panelli (2006) observes, this representation of rural community strongly informed the body of rural community studies conducted in Britain during the 1950s and 1960s (see also Frankenberg, 1966), along with the contributions of other sociologists such as Louis Wirth, which developed a dichotomization of rural and urban life that reflected the shorthand designations or rural community versus urban society. Wirth (1938), for example, identified urban life with dynamic, unstable and impersonal social relations, and rural life with a stable, integrated and stratified community, with the same people coming into contact with each other in different contexts.

Liepins (2000a) describes these early attempts at conceptualizing rural community as falling into either a structural–functionalist approach or an ethnographic perspective. Structural–functionalist accounts of rural community viewed communities as 'relatively discrete and stable phenomena with observable characteristics (structures) and demonstrable purpose' (p. 24), such as the structural characteristics identified in Tönnies's model of *gemeinschaft*. Studies of rural communities conducted from this perspective therefore set out to observe and record these anticipated structural characteristics. Studies adopting the ethnographic approach, in contrast, did not presume the existence of pre-set structural forms, but sought to document the 'real' existence and practice of community, through 'the careful recounting of "authentic" lived experiences and relationships' (p. 25). However, Liepins (also later writing as Panelli) notes that both these approaches faced difficulties in attributing either distinctive structural formations or 'authentic' practices to the concept of 'community', and as such were subjected to extensive critique and criticized as being 'descriptive, static, homogenizing, traditional, unscientific, abstractly empiricist and even pre-modern' (Panelli, 2006: 68).

Denuded of any explanatory power, 'community' fell back to being used in rural studies purely as a descriptive term, denoting 'a scale of inquiry or a loosely specified sense of social collectivity' (Liepins, 2000a: 25). It was only with the cultural turn in the late 1980s and 1990s that interest in the meaning of community was reignited in rural studies, with the emergence of a new approach that has conceptualized 'community' as a symbolic and socially constructed idea. This approach was strongly influenced by theories of symbolic interactionism in anthropology, and particularly the work of Anthony Cohen (1985), who argued that:

> the 'community' as experienced by its members does not consist in social structure or in 'the doing' of social behaviour. It inheres, rather in 'the thinking' about it. It is in this sense that we can speak of the 'community' as a symbolic, rather than a structural, construct.
>
> (Cohen, 1985: 98)

The emphasis on community as a social construction corresponded with growing interest in the plurality of different rural identities and lifestyles in the 1990s (see, for example, Cloke and Little, 1997), and hence facilitated exploration of contested communities, the connections between community and rural identity, the existence of overlapping rural communities, and the exclusionary character of some imaginings of rural community, as discussed further later in this chapter.

However, by focusing on the symbolic nature of community, the cultural approach downplays the materiality of community, as, for example, expressed in its territoriality, its institutions and meeting-places, and its practices and performances.

Accordingly, Liepins (2000a, 2000b) proposes a new conceptualization of community, which incorporates material and spatial aspects of the construction of community with cultural meanings and practices through which community is performed and reproduced. In this model, community is understood as 'a social construct about human connection that involves cultural, material and political dimensions' (Liepins, 2000a: 29), comprising the mutual interaction of meanings, practices and spaces and structures (Figure 6.1).

As Figure 6.1 shows, Liepins places people at the centre of community, as communities are made up of people, but a collection of people in itself does not make a community. Rather, a community must be ascribed with meanings, which constitute its symbolic construction. These meanings of community, though, are not free-floating, but are embodied in community spaces and structure and legitimate particular community practices. Community practices, in turn, enable the circulation and challenging of meanings – for example, the 'circulation of meanings and memories through newsletters and meetings', or 'the exchange of goods and services at a local store or health clinic' (Liepins, 2000a: 31–32) – exposing the dynamic and contested nature of rural communities. Such practices occur in the spaces and structures of the community, and indeed shape these spaces and structures, including sites such as schools, halls and bars where 'people gather in their practice of "community"' (*Ibid*: 32). Spaces and structures therefore enable the materialization of meanings and affect how practices occur. Each dimension of community, therefore, is contingent on other dimensions of community.

Liepins's model has clear methodological implications for the study of rural communities. First, there is a need to examine the texts, representations and lay discourses through which meanings of rural community are produced, reproduced and contested. Second, the ways in which community is practised and performed should be studied, paying attention

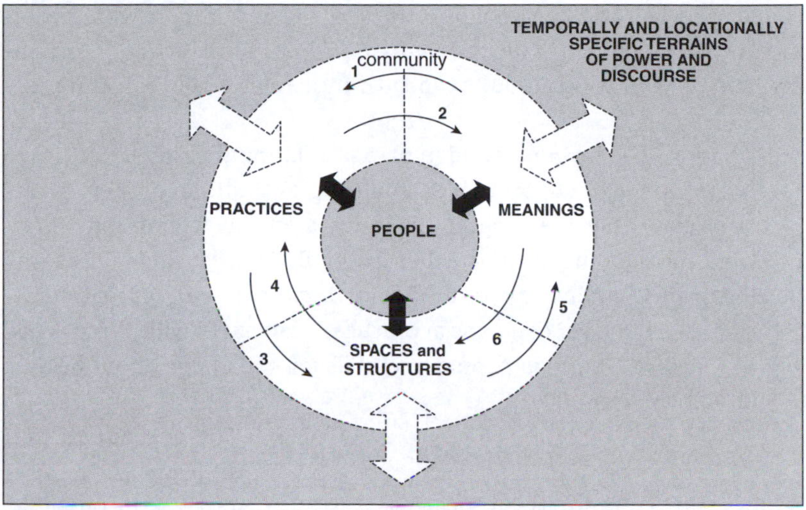

1 MEANINGS	legitimate practices
2 PRACTICES	enable the circulation and challenging of meanings
3 PRACTICES	occur in spaces and through structures, and shape those spaces and structures
4 SPACES and STRUCTURES	affect how practices can occur
5 SPACES and STRUCTURES	enable the materialization of meanings
6 MEANINGS	are embodied in spaces and structures

Figure 6.1 The constitutive components and dynamics of 'community' (after Lipeins, 2000a) (by permission of Elsevier)

both to staged events and rituals, and to everyday interactions (see Chapter 7 for more on the performance of rural community). Third, the spaces and structures of rural communities must be mapped and explored, including formal sites of community interaction such as shops and schools, but also informal spaces and structures through which community can be enacted. Moreover, Liepins's approach emphasizes the requirement for an integrated and holistic analysis of how these aspects collectively constitute community, investigated *in situ* through geographical case studies.

This is illustrated by Liepins (2000b) with case studies of three rural communities in Australia (Duaringa and Newstead) and New Zealand (Kurow). Employing participatory action research to engage local residents in the study, she demonstrates the interconnection of community meanings, practices, and spaces and structures in three towns. In Duaringa, a beef and crop farming town of fewer than 500 residents in central Queensland, for example, the meanings of community were closely associated with agriculture, as were events through which community practised, such as the Bullarama. As Liepins summarizes, the meanings and practices of community in Duaringa are mutually constitutive:

> Duaringa has been narrated in part as a 'farming community' based on services and support for surrounding agricultural properties and families. These meanings have legitimated the 'community' practices surrounding such activities as the Bullarama, Rodeo Club and the charity fund raising for the Flying Doctors. In each case, the specific voluntary work and social practices associated with these activities involve Duaringa people in mobilizing meanings about being a 'farming community'.
>
> (Liepins, 2000b: 337)

Additionally, community in Duaringa was also practised through the interaction of a largely dispersed agricultural population at keys sites such as the school and the post office, thus emphasizing the importance of the spaces and structures of community. Community practices were hence socially embedded in spaces and structures in the community, which in turn shaped how practices were enacted, not least because the spaces and structures concerned were inflected with gender and other power relations:

> In Duaringa all interviewees noted the importance of the Gold Club and the CWA [Countrywomen's Association] Hall as significant sites that provided social (and semi-'public') spaces for 'community' activities and practices ... Nevertheless, interviews and observations indicated that neither of these sites were neutral physical spaces. Indeed, the CWA Hall was recognized as a site controlled and managed by local women, while the Golf Club was more often a space that was gendered as masculine on account of the drinking practices occurring in the bar ... In both cases these 'community' sites shaped the types of activities (sport or service) and gendered practices that occurred.
>
> (Liepins, 2000b: 338)

In this way, the approach to community outlined by Liepins/Panelli reveals rural communities to be not static and stable places, as modelled in early concepts of community, but dynamic and contested groupings of people that are shaped and influenced by social and political context and power relations. As later sections of this chapter discuss, social and economic restructuring has changed rural communities by challenging the character of each of the dimensions of community meanings, community practices and community spaces and structures, and by allowing the power relations implicit in the idea of community to be contested. Yet, these developments do not mean that 'community' has become a redundant concept, but rather point to the constantly changing form of rural communities.

Rural community and belonging

Part of the adhesive that holds the various elements of community together is the notion of belonging. Belonging works in rural communities in two ways. First, it is exhibited in the sense of belonging that members of a community feel towards each other – that they share a common identity, participate in the same practices, support one another, and thus *belong to the community*. Second, belonging is also articulated in terms of a sense of belonging to place – that is the association of a particular community with a particular territorial expression (Plate 6.1).

Although Liepins/Panelli does not explicitly talk about belonging, the importance of a sense of belonging both to community and to place is evident in her case studies in Liepins (2000b). Respondents in all three localities associated community with social interaction, participating in shared events, and helping each other out; but respondents also identified sites within the area as material spaces of community, including natural features such as rivers and infrastructure such as bridges as well as the more obvious schools, halls and bars. One resident in Duaringa, for example, picked the bridge that 'everyone has to cross' (Liepins, 2000b: 335).

Similar observations were made by Neal and Walters (2008) in work in rural England on the Women's Institute (WI) and Young Farmers' Clubs (YFC) as sites through which community is practised. They show that belonging to such organizations is closely identified with belonging to a community, in that to be a member was 'the way in which to fully "be in"/belong to their particular *and* their imagined geographical place' (Neal and Walters, 2008: 285). Groups such as the WI and YFC provide a space for conviviality, which consolidates a sense of belonging, but they

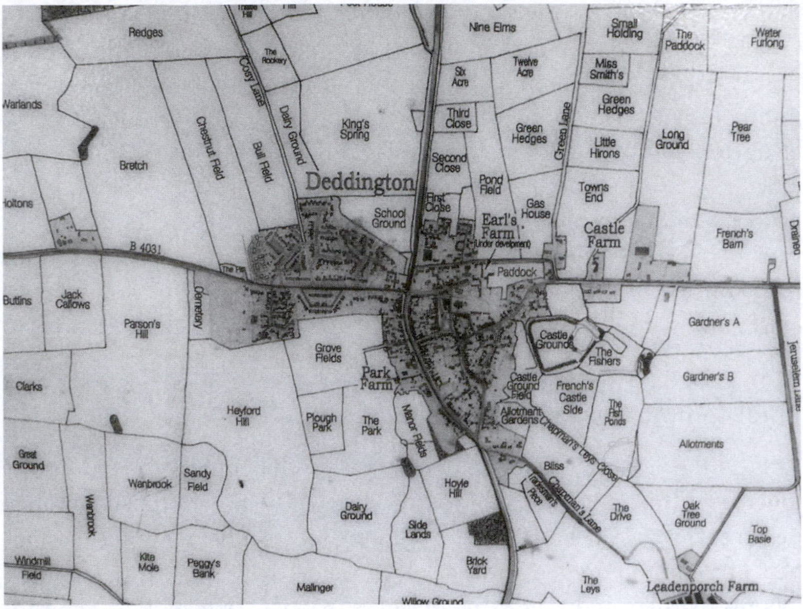

Plate 6.1 Community and territory: Deddington parish map, Oxfordshire, England (Photo: author)

also function to facilitate altruistic practices about community responsibility and care as well as social activities. Neal and Walters note that these different activities are bundled together as people talk about community-making practices – they are all about ways of belonging. Moreover, Neal and Walters suggest that the WI and YFC promote particularly *rural* forms of community practice, which serve to embed a belonging to place. With specific reference to YFC members, they note:

> What emerges from these conversations is a chain of equivalences between locationality, sociality and community. Young Farmers' Clubs create social spaces in which young people perform specifically rural behaviours ... It is in these social spaces that the production and maintenance of community occurs.
>
> (Neal and Walters, 2008: 286)

Community is here linked to place not only through the physical sites of community practice, but also through the place-rooted nature of these practices. A sense of belonging to place hence is not just about familiarity with local landmarks and recognition of territorial boundaries, but also implies a deeper knowledge of, and engagement with, the physicality of place. In a rural context, this is commonly structured by meanings of community that emphasize the centrality of agriculture. Yet, given that agriculture now employs only a small minority of the rural population in most of the global north, this association can give rise to nostalgic and defensive renderings of rural community, with rooted rural communities portrayed in opposition to the rootlessness and anonymity of modern society, as in the 'new agrarian' movement in the United States (see Box 6.1).

Box 6.1 NEW AGRARIANISM

Agrarianism has been an influential political philosophy in the United States since the founding of the union, and is commonly associated with Thomas Jefferson's assertion that, 'cultivators of the earth are most valuable citizens. They are the most vigorous, the most independent, the most virtuous, and they are tied to their country and wedded to its liberty and interests by the most lasting bonds'. The central tenets of agrarianism hold that the direct connection of the farmer to nature through the cultivation of the soil inculcates virtues such as courage, honour and moral integrity, and that rural communities rooted in the land more fully adhere to American principles such as self-reliance and independence.

Agrarian thought was popular in the early twentieth century, but the family farms that it celebrated were marginalized by the expansion of industrial agriculture. New agrarianism is a body of ideas associated with writers such as Wendell Berry, Wes Jackson and Gene Logsdon, and influenced by the earlier work of Aldo Leopold, which has developed since the 1970s. New agrarians reiterate the importance of a sense of connection between community and place, with Livingston (1996), for example, suggesting that 'a sense of community is most simply put as an awareness of simultaneously *belonging* to both a society and a place' (p. 132). However,

continued

new agrarians argue that this connection has been lost in modern society:

> the land is still more likely to be owned than known, controlled without being fully understood, and loved only for its instrumental value. Many of us remain visitors in a landscape we call home, and are estranged from the people who, in another time, would be rightly called our neighbors.
>
> (Vitek, 1996: 1)

As such, new agrarians advocate a rediscovery of the landscape and place, support for small-scale agriculture and sustainable farming methods, local food systems, and the reinvigoration of rural communities that are based around farming. New agrarianism hence resonates with progressive ideas expressed in the 'back-to-the-land' and organic movements, but its emphasis on Christian morality, ordered communities, traditional values and small-scale agrarian capitalism, and its veneration of the nostalgic model of the Anglo farmer-settler, means that it is more commonly associated with a conservative politics of community.

Further reading: Berry (2009), Vitek and Jackson (1996).

A more progressive articulation of rural community and belonging to place, however, can be found in the crofting communities of northern Scotland. As Mackenzie (2004, 2006a, 2006b) documents, the meaning of these communities is defined by the practice of crofting, a social system based on small-scale farming in which crofters have a collective and inherited right to work the land, but no right of land ownership. As such, crofting is framed by the historical legacy of the eighteenth- and nineteenth-century Highland Clearances in which crofters were displaced to permit the enclosure of the land in aristocratic estates (Mackenzie, 2004). Community identity and belonging are hence enacted both through historical references and through contemporary practice in which 'collective rights to the land are asserted on a daily basis through the rearing of sheep and management of common grazings' (Mackenzie, 2004: 283). Thus, Mackenzie observes:

> At the same time that collective rights are made visible through narratives of the past and the present, so too are collective, or community, subjectivities performed. Through the everyday practices – talking, laughing, disputing, dancing – by which the past is re-called and the present re-claimed, boundaries of identity and belonging are reworked.
>
> (Mackenzie, 2004: 285)

Mackenzie argues that the dynamic and contingent nature of these practices means that community membership is more fluid than the historical references imply, and that community belonging can be 'achieved rather than ascribed through some essentialist marker' (*Ibid*).

The necessary negotiation of community and belonging assumed new significance as crofting communities began to engage with the opportunities created by the Land Reform (Scotland) Act 2003, which gave communities a 'right to buy' the land that they occupied. Mackenzie (2006a) records the community purchase of the North Harris Estate on the Isle of Harris, which required a re-imagining of the community and its connection to the land, both in terms of establishing the rationale for the buy-out and the subsequent management of the land and in terms of defining eligibility for membership of the community trust that became the new owner of the land. What emerged was an articulation of community based on crofting principles of collective rights to the land, but recast as rights that were 'inherited *and* evolving' (Mackenzie, 2006a: 586). As Mackenzie notes, this re-articulation of community contrasts with prevailing discourses of privatization and globalization, and has become a guiding principle in the subsequent governance of the estate by the North Harris Trust (Mackenzie, 2006b).

Exclusionary rural communities

The notion of belonging can help to bind rural communities together and build 'social capital' that enables communities to act collectively, but it also implies the exclusion of people and practices that are deemed not to 'belong'. Meanings ascribed to rural communities that derive from a sense of belonging to a particular territorial place, and which emphasize continuity of residence and practice, can breed distrust and suspicion of outsiders. This can include racial and ethnic discrimination and conflicts, with groups from different ethnic backgrounds represented as not-belonging, and presenting a threat to the stability and coherence of ethnically homogeneous rural communities. Hubbard (2005), for example, describes the opposition of predominantly white rural communities in England to the construction of reception centres for asylum seekers; whilst Roma and 'New Age Traveller' populations have long been stigmatized as criminals and ostracized in many parts of Europe by rural communities that feel threatened by their mobility and different culture (Sibley, 1997; Vanderbeck, 2003). Such actions claim to protect the 'purity' of rural communities, but they are misconstrued, based on myth rather than reality. The representation of rural Europe, rural Australia and large parts of rural North America as spaces of 'whiteness' (Agyeman and Spooner, 1997; Vanderbeck, 2006), for example, ignores the indigenous peoples of America and Australia, and disguises the historic presence of non-whites in the countryside of rural Europe (Bressey, 2009).

Ethnic minorities living in rural communities frequently experience marginalization and victimization (Chakraborti and Garland, 2004). In extreme cases they can be subjected to racial abuse and racist attacks, but more commonly they are the victims of 'covert racism', comprising 'mechanisms, assumptions, inflections and orthodoxies which serve to deny people of colour any distinct cultural identity in rural settings, presenting them with lifestyle choices involving a denial of ethnic identity so as to "fit in"' (Cloke, 2004: 30). Ethnic minority residents can find themselves excluded, directly or indirectly, from the social networks and practices through which community is practised, as Tyler (2006) describes for Asian households in villages in Leicestershire in England, signalling that they are not perceived to fully belong.

As such, the practices of community can also be exclusionary, especially where these are associated with religion or cultural practices such as drinking alcohol. As Garland and Chakraborti (2004) observe in the context of rural England, 'the "customary" visit to the village pub, especially on a Sunday, also causes problems for those whose faith dictates that they should not drink alcohol' (p. 127). Furthermore, rural community practices and events are also frequently inscribed with gender stereotypes

(Hughes, 1997; Liepins, 2000a; Neal and Walters, 2008) and norms of heterosexual behaviour (Little, 2003). Behaviours that do not conform to these expectations cannot be accommodated within the community, and attempts to establish alternative spaces in which alternative community practices can be articulated may become the focus of conflict over whether or not they 'belong' in the community, as Gorman-Murray *et al.* (2008) describe in the case of a gay and lesbian festival in the small town of Daylesford, Australia.

Control over the spaces and structures of community life also facilitates regulation of behaviours within the community, as well as the exclusion of those deemed not to belong. Neal and Walters (2007) comment that the everyday practices that sustain community sensibility involve 'a heavy reliance on the notion of neighbour knowledge and the importance of the informal and formal processes of watching and surveillance' (p. 254). They note that surveillance is often constructed positively, as caring watchfulness, 'watching over those identified as being captured by community boundaries and watching out for those that are outside of these' (p. 255), and structures of surveillance such as Neighbourhood Watch schemes may themselves become practices of community-making.

Surveillance can also function to enforce conformity within rural communities and to police activities that are a perceived to be disruptive to the rural idyll, rather than necessarily illegal or threatening. The pressure to conform and the regulation of behaviour can be particularly felt by young people, whose everyday activities such as gathering in public spaces or skateboarding on footpaths, may be perceived as disruptive and threatening by other community members (Panelli *et al.*, 2002; Rye, 2006). However, Leyshon (2008) presents a more complex picture of youth belonging, noting that young people tend to reproduce the same discourses of rural community as adults, and that they frequently appreciate the safety and mutual help that follows from a strong sense of community. Moreover, young people (and adults) in rural communities are adept at testing the boundaries of regulation, and at finding marginal spaces that can be colonized for their own use, or where deviant practices (smoking, drinking, drug-taking, non-conforming sexual behaviour) can be performed 'out of view' (Neal and Walters, 2007; Panelli *et al.*, 2002). In this way, young people are active participants in the construction of rural communities.

Rural poverty and deprivation

Negotiating practices of belonging and inclusion in rural communities can be particularly sensitive when it comes to questions of poverty and deprivation. In most of the world, poverty is perceived to be more prevalent and persistent in rural regions – indeed, rural poverty is a key push factor in rural-to-urban migration in the global south (Lynch, 2005). In more prosperous countries of the global north, however, the perception tends to be reversed, with Milbourne (2004) observing that 'the city has assumed the status of the "natural" place of poverty' (p. 13). This reflects the more visible concentration of poverty in urban areas, and also the power of the rural idyll myth. In actuality, the rate of poverty in the United States is higher in rural counties (14.8 per cent in 2000) than in metropolitan counties (11.9 per cent) (Lichter and Johnson, 2007), and substantial levels of poverty were identified in several rural case study areas in England and Wales by research in the 1990s (Milbourne, 2004).

Such is the strength of the perception, however, that even many rural residents who would objectively be categorized as living in poverty deny the existence of deprivation in their communities. For example, in an area of Nottinghamshire in England, where nearly four in ten households fell beneath a commonly accepted measure of the poverty line, only 21 per cent of residents and only 15 per cent of residents in poor households reported that there was deprivation in their area (see Table 6.1) (Milbourne, 2004). Equally, elite groups in rural communities can adopt narratives that explain away the presence of poverty, constructing it as a matter of choice or personal failure. Lawson *et al.* (2008), drawing on case studies in Idaho and Montana, report that:

> The White poor are frequently explained as choosing their lifestyle, as anti-establishment and rejecting assistance, as not wanting the (farm) work that is available. Latino poor are framed as undeserving, criminal and/or threatening, as happy for backbreaking low-wage work and as needy and draining public resources. These understandings of poor subjects obscure the failures of pro-market policies and reinforce community cohesion through constructions of the poor as deviant and failing.
>
> (Lawson *et al.*, 2008: 750)

Table 6.1 Poverty rates and perceptions of deprivation in selected case study areas in rural England, 1990–1 (after Cloke *et al.*, 1995; Milbourne, 2004)

Study area	**Households with incomes of less than 140% of Income Support entitlement (%)**	**Respondents reporting the presence of deprivation in the local area**	
		All respondents (%)	**Respondents in poor households (%)**
Nottinghamshire	39.2	21.0	15.0
Devon	34.4	43.9	45.5
Essex	29.5	29.3	38.5
North Yorkshire	22.0	46.2	22.2
Northamptonshire	14.8	25.1	25.0
Cheshire	12.8	32.1	20.0

At the same time, elite respondents in the study reproduced imaginaries of place that erased poverty, representing communities in ways that left no space for poverty, describing true local residents as rugged frontier individuals who would reject assistance, and suggesting that the poor were 'just transients moving through' (Lawson *et al.*, 2008: 750). Representations such as these construct a moral framework that can constrain the coping strategies available to rural residents experiencing poverty, as Sherman (2006) demonstrates for a small community in California. As community residents are subjected to considerable social pressure to uphold its cultural norms and to 'act in mainstream ways' (p. 907), they make individual decisions about coping strategies based on the perceived moral capital of different options. This led some poorer residents to reject strategies such as accepting welfare payments in favour of more 'morally acceptable' routes such as taking ad hoc menial paid work and growing their own food.

In countries where recognition of rural poverty is more widespread, differences of interpretation can still exist over the meaning of poverty between rural residents and external actors, such as government agencies and aid organizations. Kadigi *et al.* (2007), for example, report that when poor rural residents in Tanzania were asked what poverty meant to them, they emphasized not income, but 'intangible assets' such as health, literacy, and a sense of their voice relative to other members of the community. Shubin (2007) similarly argues that rural poverty in Russia should be understood as a relational or networked condition:

> Poverty can be seen as a fusion of different events and it unfolds through different connections or misconnections where people are involved or not involved. Poverty is loneliness and isolation, but it is also involvement in the systems of help because of people's knowledge of local symbolic practices such as collecting firewood or swapping agricultural products. Poverty is not neatly categorized simply on the basis of income or remoteness, but is constructed within complex webs of relationships in the village. Most poverty experiences are networked: both material items such as bread and pensions, and non-material services such as medical help, assistance from *dachniki* [second home owners] and the lack of help from [family], are reproduced within the local webs of communication.
>
> (Shubin, 2007: 594)

In this way, deprived rural residents develop coping strategies that draw on relations within the community, in turn shaping individual self-perceptions of what it means to be poor in rural Russia and informing collective discourses of the 'deserving' and 'undeserving' poor that are based on and fit with a dominant agriculture-centred representation of poverty.

RE-MAKING RURAL COMMUNITIES

As the previous section has discussed, there is no such thing as a stable, static and homogeneous rural community, as early theories of community suggested. Indeed, it can be argued that there are many overlapping and plural rural communities co-existing across the same rural territory, sometimes sharing spaces of interactions and practices, and sometimes competing in their claims to place. Rural communities are hence dynamic and contingent, and are constantly evolving in the context of social and economic restructuring. In particular, the constitution of rural communities has been challenged by the increased mobility of rural society, which has at least three manifestations. First, increased mobility has been associated with the out-migration of people from rural communities, depopulating rural areas and threatening the viability of community structures. Second, in parts of the global north, increased mobility is associated with the opposite trend, of counterurbanization and in-migration to rural communities. Third, members of rural communities are themselves more mobile. On a very short time scale, this mobility is expressed in commuting to work and travel for shopping and leisure. On a longer time scale, it includes the mobility of some part-time residents between different properties, and temporary migration for work, both in to and out of rural communities. Collectively, these processes are continuously re-making rural communities, as this section investigates.

Out-migration and the dilemma of rural belonging

Out-migration is the dominant population trend for the majority of rural communities around the world. In the global south, rural-to-urban migration is fuelled by push factors including depeasantization, poverty, famine, war and natural disasters in rural regions, as well as by the pull factors represented by prospects of better employment opportunities, education and healthcare in cities (Lynch, 2005). Similarly, in the global north, in spite of the considerable attention devoted to counterurbanization (discussed below), out-migration still prevails in many regions including the Great Plains of the United States, the Australian interior, most of rural Eastern Europe, and peripheral districts of France, Italy and Scandinavia. The consequence of out-migration is to deplete the critical mass of rural communities. A lower population means fewer people to participate in the events and rituals through which community is practised, whilst insufficient custom may lead to the closure of shops, schools and other facilities that had formed sites of community interaction (Stockdale, 2004). With little new development, the meanings attributed to community might become even more entrenched in nostalgia, engendering a defensive and pessimistic outlook.

The out-migration of young people can be particularly damaging, skewing the demographic profile of the community and removing the most economically active section of the population (Stockdale, 2004). Even in rural Britain, where counterurbanization has been dominant for 30 years, there is a net out-migration of young people from rural communities. In some cases, this migration reflects the appeal of the city and the desire of young people to escape the perceived dullness of rural life

(Rye, 2006). However, in many cases young people are compelled to leave rural communities for education, employment or to find affordable housing, and studies have shown that their attitudes towards migration are often ambivalent and reveal the complexities of belonging in rural communities.

Research in England and Norway has demonstrated the strong sense of belonging felt by many rural young people towards their community (Leyshon, 2008; Rye, 2006), and as such decisions about leaving can involve a difficult choice between emotional attachment and economic opportunity. Analyses in Australia and Iceland show that young people with the strongest sense of belonging to the community, and those raised in the community, are less likely to leave, but that opportunities for desired occupations are the strongest predictor of migration intentions (Bjarnason and Thorlindsson, 2006; Pretty *et al.*, 2006). The migration decision can also be influenced by cultural expectations, with Ni Laoire (2001) noting that in rural Ireland, for example, there is 'a certain stigma attached to staying', and that 'the association of migration with heroism contributes to the devaluation of staying' (p. 224). Staying behind hence means risking being perceived as a failure within the community.

There is therefore a strong narrative of improvement associated with out-migration from rural communities, which Wiborg (2004) presents as a 'class journey' from a disadvantaged rural background to a different lifestyle. Paradoxically, the ambition for improvement might be heightened by the strength of belonging to the home community. Baker and Brown (2008), in a study of first-generation university students from rural Wales, for example, point to the importance of an 'aspirational habitus', that was 'made possible by the isolation of the community' (p. 67), with community identity and educational aspiration reinforced through chapel, Sunday school, family and village school.

Corbett (2007a), in a study of education in coastal Nova Scotia, Canada, similarly argues that rural schooling develops educational mobility capital that distances young people from their homes and communities. They are in essence, 'learning to leave'. However, Corbett also argues that the emphasis on educational aspiration and mobility contributes to the alienation from formal education of those who choose to stay. This in turn reinforces a gender gap in rural communities, with women more likely to leave than men, in part because they perform better in formal education (Corbett, 2007b), and in part because of the masculinist culture of many rural communities (Ni Laoire, 2001). As Ni Laoire (2001) observes in rural Ireland, 'evidence suggests the persistence of a dichotomy between the necessary spatial mobility of young women and the spatial immobility of many young men' (p. 224). Moreover, Ni Laoire argues that the disproportionate rates of out-migration, combined with the stigma associated with staying behind, and the decline of primary industries, are contributing to a crisis of masculinity in regions such as rural Ireland, as evidenced in part by a rise in male suicide rates.

In-migration and the contestation of place

Rural migration is two-way, and in the early 1970s rural geographers in the United States noted that for the first time in decades, migration into rural areas exceeded migration from rural areas. This new situation of counterurbanization was subsequently documented in many parts of the global north, including Canada, Britain, Europe, Australia and New Zealand. In England, for example, the population of rural districts increased by 12.4 per cent between 1981 and 2001, compared with an increase of only 2.4 per cent in the urban population (Woods, 2005a). Over the decades since 1970, the relative balance of urbanization and counterurbanization has oscillated in several countries, including the United States, and recent analyses have proved counterurbanization to be a more complex and differentiated process than early accounts suggested (see Box 6.2). Nonetheless, in-migration has become a major factor impacting on many rural communities. These include certain localities in South and Central America and South East Asia that have been targeted by international amenity migrants (Moss, 2006). The San Antonio district in central Costa Rica, for example, has experienced a significant influx of amenity migrants since the 1970s, such that by 2000 a quarter of its population comprised foreign in-migrants, drawn from 81 different countries (Chaverri , 2006).

Migration into rural communities for economic motivations are at least a partial factor for most in-migrants. However, recreational and lifestyle opportunities are important for many migrants, with destination localities selected that appear to correspond with popular discourses of rurality (Plate 6.2). Halfacree (1994) demonstrated that the narratives of in-migrants to rural communities in England strongly emphasized the deliberate selection of a rural setting, often described in language that

Box 6.2 COUNTERURBANIZATION

Counterurbanization refers to a trend of population change in which rural population growth exceeds that of the urban population, such that the balance of population in a nation or region shifts towards rural areas. It represents a reversal of the historic trend of urbanization in which the rural population is depleted by the growth of urban centres – still the dominant trend in most of the world. The label of 'counterurbanization' was first applied by Berry (1976) to the 'population turnaround' observed in the United States in the early 1970s as rural population growth started to out-strip urban growth, and has subsequently been identified in other parts of the global north.

Counterurbanization comprises a range of components including urban to rural migration, and also rural-to-rural migration, immigration and natural population evolution. It includes the 'decentralization' of populations from cities to peri-urban districts, and also 'population deconcentration', or long-distance migration into peripheral rural regions. However, counterurbanization does not necessarily mean urban populations are decreasing. In Britain and the United States, for example, both urban and rural populations increased during the 1990s, but the rural population expanded most rapidly. Counterurbanization also encompasses a variety of motives for migration, including economic and labour migration, retirement migration, return migration and amenity migration.

Kontuly (1998) accordingly identifies diverse drivers of counterurbanization including economic cyclical factors, economic structural factors, spatial and environmental factors such as the amenity appeal of rural areas, socio-economic and socio-cultural factors such as changes in residential preferences, government policies and technological innovations. In essence, though, the preconditions for counterurbanization are an economy that is not dependent on place-tied industries, an affluent and mobile urban population and a developed consumer society, and rural communities that can provide an equivalent standard of living to urban areas. As such, it is a feature of advanced industrialized societies that have started to de-industrialize.

The strength of counterurbanization in Britain and, to a lesser extent, the United States has given it considerable prominence in Anglo-American rural geography. Recently, however, the evidence for counterurbanization has been subjected to a more critical re-appraisal (e.g. Halfacree, 2008; Mitchell, 2004; Smith, 2007). First, critiques have noted that the pattern of counterurbanization is neither consistent nor continuous. With the notable exception of Britain, countries in Europe and North America have experienced periods of counterurbanization interspersed with periods of urbanization over the last 30 years (Kontuly, 1998; Woods, 2005a). In several countries, counterurbanization has reflected particular regional dynamics – in Australia population growth on the rural east coast has contrasted with depopulation in the rural interior, whilst in the United States rural population growth in the south and west contrasts with depopulation in the rural Mid-West (Woods, 2005a). Even in counterurbanizing regions there can be local pockets of depopulation, and local trends can reveal a consolidation of population in small towns – as Walford (2007) shows in rural Mid-Wales.

Second, the importance of the 'rural idyll' in attracting migrants to the countryside has been questioned. Although emphasized in many early studies, the significance of the rural idyll has been downplayed in more recent analyses that have instead prioritized economic factors and family ties in explaining migration decisions (Stockdale, 2004). Similarly, third, the class character of counterurbanization has been re-assessed. Early studies tended to associate counterurbanization with the middle class, especially the 'service class' of professionals and managers (Cloke *et al.*, 1995). More recent work suggests that middle-class ex-urbanites constitute a minority fraction within counterurbanization, alongside economic migrants filling lower-order jobs (including migrant workers), 'welfare migrants' claiming benefits from the state, retirement migrants motivated by cost of living as well as rural amenity, health migrants and return migrants. Migrants motivated by the quest for rural lifestyles are also a differentiated group, including not only stereotypical middle-class

continued

idyll-seekers, but also younger amenity migrants pursuing recreational opportunities such as surfing or mountain biking, as well as purchasers of small-holders aiming to go 'back-to-the-land' (Halfacree, 2007). Moreover, counterurbanization includes the formation of various 'intentional communities', each embodying a different take on rural life, from sustainable communities and eco-villages, to religious communities, to communities of lesbian women (Meijering *et al.*, 2007).

Counterurbanization is hence a more complex and differentiated phenomenon than has often been acknowledged in the rural studies literature. Mitchell (2004) suggests categorization into three sub-processes of 'ex-urbanization', 'displaced urbanization' and 'anti-urbanization', defined by spatial and economic dynamics and household motivation. However, even this model fails to capture the full diversity of population dynamics operating through rural areas.

Further reading: Boyle and Halfacree (1998), Halfacree (2008), Mitchell (2004).

resonated with the discourse of the 'rural idyll'. In Australia, meanwhile, migration to rural coastal communities has been dubbed the 'sea change' movement, reflecting its association with a desired lifestyle change (Burnley and Murphy, 2004). As Burnley and Murphy (2004) record, a significant minority of such migrants cited the ambition of moving closer to nature, or living in a rural area, as key reasons for migration. Thus, migration into a rural community can be driven by the aspiration to adopt a perceived rural lifestyle.

In-migration in substantial numbers necessarily changes rural communities, as is vividly evidenced in Bell's (1994) classic ethnographic study of the pseudonymous village of 'Childerley' in southern England. The village sits on the edge of the London commuter belt, and as such has experienced considerable in-migration, yet as Bell describes, implicit boundaries were imagined in the community to divide 'local' residents from the ex-urban in-migrants:

Plate 6.2 New house-building at Wanaka, New Zealand, a popular amenity migration destination (Photo: author)

> The residents refer to themselves as true villagers, country cousins, country bumpkins, locals, country girls, countrymen bred and born, Hampshire hogs, salt of the earth, real countrywomen, village people – as well as what I have adopted as the broadest term, country people. Others they describe as city dwellers, bloody townies, Londoners, yuppies, city slickers, city-ites, outsiders, foreigners, day-trippers, town people, as well as city people ... The phrase city people, in the views of most Childerleyans, fits many current residents of the village.
>
> (Bell, 1994: 101)

Such rhetoric turns a geographical differentiation into a cultural differentiation. It is possible for an in-migrant to win acceptance as a local, yet this achievement is not based on residence time, but on learning and adopting 'country ways'. Local rural identity is based on culture, on knowing and understanding nature and how to use it, and on knowing the landscape to which the community belongs. These attributes are emphasized not as an exclusionary test, but because they relate to the meaning of community and to people's sense of belonging to place. Harper (1988), in research elsewhere in England, similarly observed that whilst references to sites in the community by locally born residents were rich in historical resonance, mentioning former residents and past events, in-migrants to the community tended to describe the village landscape and environment in comparison with the place where they had lived before.

Boundaries within communities can also be erected by in-migrants through their choices about how and where to live and socialise. In some cases, in-migrants eschew the established structures and practices of the community, in favour of building social networks that reflect a community of interest over the community of place. Bell, for instance, quotes one in-migrant who acknowledges that 'I suppose some people make friends with their next door neighbours because it's their next door neighbours. I think we tend to only make friends with them if they were similar sorts of people to ourselves anyway' (Bell, 1994: 98). Similarly, Salamon (2003), in a study of newcomers to rural towns in Illinois, observes that in-migrants have weaker social ties in the community, and are less likely to do something because it is part of belonging to the community. She notes, for example, that newcomers may choose to go to church somewhere else, that 'choice of church is linked more to theology than to community membership and status' (p. 16), and that middle-class in-migrants tend to segregate their children rather than subscribe to practices of shared responsibility and watchfulness over all children in the community. Moreover, with specific reference to one case study community, she argues that:

> Newcomers engage in community activities narrowly, only in those institutions from which they see direct personal benefit – the schools or a church. Newcomers enjoy Prairieview's rural ambience and the safety and security that its size and homogeneity provide, but not those qualities that made it special to oldtimers ... When a town serves only as a residence space, the inhabitants do not look to the town to provide a unique place identity or social status, as oldtimers did.
>
> (Salamon, 2003: 179–80)

Salamon's description of Prairieview presents a rather instrumentalist view of in-migrants' engagement with place. As noted above, the perceived sense of community found in the countryside and the desire to participate in a rural lifestyle can also be important motivations for in-migrants, and there are many who adhere to a 'move-in and join-in' philosophy (Cloke *et al.*, 1998). There are many examples of successful integration, but also many examples of in-migrants creating or colonizing structures and spaces of community that exist in parallel to those of the established population. In essence, there are plural rural communities co-existing in the same territorial space, and inevitably conflicts arise over the use and appearance of the landscape, and over the meaning of the community. Local struggles have developed over issues such as new buildings, noise and disruption from agricultural activities, access to footpaths, and 'improvements' such as street lighting, and in-migrants have been motivated to stand for election to local government bodies to represent the views of their 'community' (Woods, 2005b).

However, the studies by Bell (1994) in England and Salamon (2003) in the United States both also show that differences between in-migrants and the established local population may have more to do with class than with geographical background. Indeed, the rhetoric of local/newcomer tensions may serve to obscure class conflict in rural areas, hence maintaining the pretence of a singular, class-less rural community. In reality, counterurbanization has changed the class composition of many rural communities. In countries such as Britain, migration to rural areas has been particularly associated with the professional middle classes (although see Box 6.2 for a critique of this assumption). The purchasing power of middle-class in-migrants means that they have been able to out-price working-class rural residents in rural property markets. The resulting displacement of rural working classes may be presented as a process of gentrification, which has material effects in changing rural communities (Phillips, 2002). Modifications are frequently made to housing, either to conform to certain perceptions of rural living, or to facilitate forms of urban living, expectations that are also appealed to in speculative property developments such as barn conversions or gated communities. At the same time, in-migrants can mobilize politically to protect the middle-class character of their adopted communities, opposing new housing developments that would increase supply and potentially reduce property prices (Murdoch and Marsden, 1994).

Not all in-migrants to rural communities are middle-class ex-urbanites, and the significance of return migration has been increasingly recognized in recent studies (e.g. Falk *et al.*, 2004; Ni Laoire, 2007; Stockdale, 2006). Wiborg's (2004) research with students from rural regions in Norway shows that they retain a sense of belonging to their home community, but that the nature of this relationship changes such that the rural represents the place that they have come from, as opposed to a community in which they actively participate. Many people who migrate from rural communities for education or employment intend to return, but may be prevented from doing so by a lack of appropriate jobs, limited affordable housing or new relationships and family circumstances. As such, return migration tends to occur later in a working life, or for retirement (Jauhiainen, 2009). Stockdale (2006) argues that return migration is essential for rural development, bringing in skills and capital, but not everyone returns because they have succeeded elsewhere. Indeed, fear of being branded a failure is a factor that can prevent younger out-migrants from returning home (Stockdale, 2006).

The decision to move back to a rural community can be complex and reflect several inter-connected motivations. Research in rural Ireland, for example, suggests that primary reasons for returning include being closer to family, and seeking a particular remembered lifestyle. This latter motive can be inflected with ideas about rurality, such that return migrants may 'express an active desire to live in the countryside, referring to a slow pace of life, safety, a good place to bring up children, and in general a better quality of life than is possible in an urban area' (Ni Laiore, 2007: 337). Yet Ni Laiore also reports that the experiences of return migrants are often at variance with their expectations. Not only can semi-mythologized ideas of rural life fail to translate into reality, but return migrants often experience a 'culture shock' in swapping urban anonymity for known histories and identities, and, paradoxically, loneliness. Return migrants can find that they occupy an ambiguous position as 'insider-outsiders', connected to the community by previous history, but not entirely belonging. For this reason some return migrants fail to settle and eventually move on again, as Stockdale (2006) observes in rural Scotland.

Mobile rural communities

The challenge of mobility to settled notions of rural community comes not only from intensified migration to and from rural areas, but also from the increased mobility of rural residents themselves. Commuting for work is now commonplace in rural regions, especially in the global north, and rural residents also routinely travel outside their communities for leisure and shopping. Some rural residents will be absent for longer periods for education or employment. Some will spend part of the year away at second homes, or on extended holidays. These forms of 'everyday mobility' tend to be taken for granted and are rarely problematized (Gerrard, 2008), but they can have a disruptive effect on the established structures and practices of a community by reducing the critical mass of people participating.

Moreover, many rural communities play host to transitory residents, whose presence in the locality is short term, seasonal or periodic, including second home owners and migrant workers. The ownership of a 'second home' or 'holiday home' in the countryside has long been popular in Scandinavia and New Zealand, but numbers of rural second home owners increased during the 1980s and 1990s in countries such as Britain, Canada, Germany, Italy, Spain and the United States (Gallent *et al.*, 2005; Halseth and Rosenberg, 1995), in line with the fashion for rural amenity consumption (see Chapter 4). As such, second homes are often concentrated in coastal and mountain areas, and in landscapes conforming to the ideal of the 'rural idyll' close to major cities. In particularly sought-after locations, second homes can comprise over a third of the local housing stock, thus adding to house price inflation, squeezing the availability of affordable housing, and reducing the permanent population that is present to support village services such as shops and schools (Gallent *et al.*, 2005). Additionally, as cheap air travel has meant that 'the average "acceptable" distance between first and second home has been increasing and many second homes are owned by non-nationals' (Schmied, 2005: 153), cultural tensions can develop between part-time foreign residents and full-time national residents.

Yet, second home owners are different to tourists. They have invested in a place and return regularly, frequently aspire to be part of the community, and develop a sense of belonging to the locality of their second home. Indeed, research by Stedman (2006) on second home owners in northern Wisconsin concluded that they 'have extensive experience in the area, important social relationships, and exhibit higher levels of place attachment than year-round residents' (p. 201). However, the nature of attachment contrasts with that of permanent residents, relating more to the landscape, environmental qualities and the sense of escape from day-to-day cares, rather than the social networks and community meanings that are central to the sense of belonging of year-round residents (Stedman, 2006).

The second key group of transitory residents in many rural regions, especially in the global north, is migrant workers. Migrant labour has always been an important element in agriculture in California (see Chapter 3), but the practice has become more widespread as social and economic transformations have created a stratum of low-grade jobs in industries such as agriculture, food processing, construction and tourism that often cannot be filled by local rural labour. As such, a number of regional trans-national circuits of labour migration have emerged: Latin American migrant workers in the United States; Eastern Europeans in Britain, Ireland and Scandinavia; North Africans in Italy and Spain; Albanians in Greece; Filipinos in Taiwan; Pacific Islanders in New Zealand; and so on. However, blanket reference to 'migrant workers' disguises a highly complex and differentiated set of migration dynamics, with individuals moving for different reasons, with different intended periods of stay, working with different employment conditions, and having different experiences of engagement with local rural communities.

Seasonal agricultural workers, for instance, are frequently subjected to poor levels of pay and employment conditions (Rogaly, 2006; Rye and Andrzejewska, 2010), and housed by employers in basic accommodation away from rural settlements. As such, they are structurally separated from rural communities. However, an increasing number of migrant workers outside agriculture live and work in rural towns and villages, and thus are a visible, if challenging, presence in communities. In some places, migrant workers have faced racist attacks and discrimination from elements in the local population, but more commonly there is a broad acceptance of the contribution made by migrants to the rural economy. For their part, migrant workers often consider rural areas to offer greater safety than cities. It is this uneasy understanding that Torres *et al.* (2006) refer to as the 'silent bargain' in the rural south of the United States:

> For employers, the social and cultural difference that Latinos represent is subject to discipline and control. For Latinos, the *tranquilidad* of the rural experience becomes an acceptable trade-off for serving as a low-paid workforce subject to exploitation.
>
> (Torres *et al.*, 2006: 38)

The delicate balance of this accommodation is tested as the migrant population expands and becomes more settled. Latino migration to the rural United States has changed in character over the last two decades, as temporary seasonal migration has been replaced by longer-term migration, farm work has been supplemented by involvement in other sectors, and migrants have pushed in a wider geographical range of rural areas in the South, West and Mid-West (Millard and Chapa, 2004; Smith and Furuseth, 2006). Like other in-migrants to rural communities, Latino migrants negotiate their engagement with the existing community, and the formation of sense of belonging, but outcomes vary shaped by factors including the class and industrial structure of the locality, and histories of immigration (Nelson and Hiemstra, 2008).

In places such as Leadville, Colorado, migrants remain marginalized as a parallel structure of community facilities – shops, churches, hair salons – has been established by Latinos, but only serving the Latino population. Where Latino and non-Latino residents use the same facilities, they do so at different times of day, reflecting different patterns and conditions of employment. Thus, 'interactions between immigrant and non-immigrant residents can generally be characterized by paucity of substantive contact – the time-space geographies of each group rarely intersect' (Nelson and Hiemstra, 2008: 324). In other places, such as Woodburn, Oregon, the political organization of Latino migrant workers has afforded them a greater visibility in the community, such that 'low-wage and racialized immigrants arriving in the mid-1980s and beyond could more easily develop a sense of belonging that included "Mexicanness" and difference' (p. 336). Whilst the spaces and structures of community might continue to be segregated, the status of negotiated coexistence potentially allows for the development of a form of 'rural cosmopolitanism' (Torres *et al.*, 2006).

One critical aspect of a rural cosmopolitanism would be the stretching of community identities over space. Skaptadóttir and Wojtynska (2008), for example, draw on Appadurai (1996) to describe the presence of Polish migrant workers in Icelandic fishing communities as creating 'post-national zones', or 'spaces where people of different origins meet, compete and negotiate their place' (Skaptadóttir and Wojtynska, 2008: 119). Significantly, Skaptadóttir and Wojtynska also describe the Polish migrants as following 'dichotomized lives', 'working in Iceland in order to create a better life back home' (p. 124); and as maintaining a 'bifocal' view, in which 'some people claim to have two homes, both "here" and "there"' (p. 123).

Rural regions are also major sources for migrant labour, and as such the stretching of community can also be observed from the 'supply' end of the chain. Migrant workers from rural regions in both domestic cities and foreign countries keep a sense of belonging to their home community, and can continue to participate in the community economically, socially and culturally. Englund (2002), for instance, observes that rural migrants in Lilongwe, the capital city of Malawi, 'see their stay in town through the prism of their rural aspirations' (p. 153), with nearly 90 per cent intending to return home at some point. Moreover, he suggests that the migrants feel spiritually connected to rural communities, with Malawian traditions of witchcraft used to explain the inter-relation of happenings in rural and urban settings, such that rural and urban spaces overlap in the migrants' lives. In this way, Englund argues, 'the domain of the rural, both as the object of moral imagination and as a geographical site, is constantly remade in relation to what migrants achieve and fail to achieve during their stays in town' (p. 153).

Velayutham and Wise (2005) similarly posit the notion of the 'translocal village' to describe 'a particular form of moral community based around village-scale, place-oriented familial and neighbourly ties that have subsequently expanded across extended space' (pp. 38–39). With reference to a case study of migrants from a Tamil village in south India to Singapore, Velayutham and Wise demonstrate that the 'translocal village' encompasses both economic and moral ties to the home community (e.g. sending remittances, helping other community members to migrate), and 'the replication, or mirroring, of village activities in the place of settlement' (p. 32). The social and political structure of the village is hence replicated in the migrant population in Singapore, and events and rites of passage in the village are also marked in Singapore. Accordingly, 'the social field of the translocal village is reproduced as much through the translocalizing of affective regimes emerging from the village, while rituals, sentiment and affect are means of inscribing locality onto bodies, creating a sense of boundary across extended space' (*Ibid*: 41).

MORE-THAN-HUMAN RURAL LIVING

Liepins (2000a) put people at the centre of her model of community, and it is primarily in terms of human communities that we think about life in the countryside. However, rural space is shared with non-human life, including both animals and vegetation, which also have ways of 'living' in the rural. Indeed, the relationship between human and non-human, or people and nature, is heavily stressed in discourses of rurality. The notions of 'countryism' employed by long-term residents in Bell's (1994) study village of Childerley, for example, suggested that 'true' rural people understood nature, and knew how to live alongside nature, in a way that non-rural incomers did not. As Jones (2003) comments, 'animals are central to how the rural is constructed in both imaginative and material terms' (p. 283), whilst Buller (2004) similarly observes that, 'society's relationship to animals in general has largely grown out of the set of activities and endeavours that are traditionally associated with the countryside, from animal breeding and domestication to the aesthetic appreciation of a gentle, ordered and unthreatening nature' (p. 131).

Traditional rural discourses of nature are, however, framed by a taxonomy of flora and fauna that both dictates the appropriate forms of relationship between humans and non-humans, and establishes the spaces to which particular forms of life belong (Buller, 2004; Jones, 2003). Certain animals are accepted as pets and allowed to share the domestic spaces of people, whilst certain plants are similarly designated as 'garden plants' and may feature as such in imaginings of the rural idyll, as in the cliché of roses around the cottage door, or the notion of the 'country garden'. Other plants and animals are defined by their contribution to agriculture, in some cases to the extent that they only exist in their present, modern form because of selective breeding to maximize specific attributes for agriculture. Livestock are animals that are solely defined by the domestication for agriculture and which have a particular place within the countryside in farm units and on farmland. Crops similarly are defined by their agricultural function and spatially ordered in fields and orchards. Beyond these spaces of domesticated nature are found the 'wild things' (Buller, 2004): animals and plants that have not been tamed or controlled by humans. Some wildlife is regarded in traditional rural discourses as benign, or even as valuable, other wildlife is constructed as threatening to human life, livestock or crops.

The orderly functioning of the traditional countryside hence rests upon the maintenance of spatial boundaries between different types of nature. Trouble occurs when these spatial boundaries are transgressed: when livestock escapes from fields, or, more commonly, when wild nature invades the spaces of the farm, as weeds in crop fields, or predatory animals attacking livestock. The fox, for example, becomes a threat to farming when it penetrates the spatial sanctuary of the chicken coop (Woods, 2000). In traditional rural discourses of nature, therefore, wildlife needs to be controlled and regulated, notably through the hunting of predatory animals. The ritual practice of such control can become important performances in the constitution of rural identity and rural communities (see Chapter 7).

However, as Buller (2004) notes, both the discursive and the material construction of human relations with nature have evolved over time. Historically this has meant the taming of rural nature, both as subjugation for agricultural production and to produce a 'safe' countryside for human activity, at least in regions such as western Europe:

> Domestication has been extended far beyond the traditional beasts of burden to include the entire faunistic composition of rural space and, more recently still, its gene pool. The countryside has become largely devoid of the larger predatory species who not only thrive on the very animals humanity has painstakingly bred and engineered, but also compete with hunters for the few remaining species deemed interesting to chase and kill. It is grazed countryside. Species are protected certainly but in wildlife parks, zoos and special zones where lions and tigers are as common, if not more so, than formerly indigenous species such as the wolf and the otter. This is a safe countryside where humanity nurtures and is, in return, nurtured by an accessible, appropriated and unthreateningly recognisable nature.
>
> (Buller, 2004: 132)

Moreover, as the means by which an increasingly urbanized population engages with nature is largely mediated by television wildlife documentaries, films and literature, nature of all types has become anthropomorphized. This, in turn, has fed a range of new discourses of animal rights, animal welfare and environmental protection, which promote a politics of conservation that can conflict with traditional rural discourses of nature, and challenge the meaning and practices of indigenous rural communities.

McGregor (2005), for example, describes conflict over the conservation or control of Nile crocodiles in Lake Kariba in Zimbabwe. The crocodiles are dangerous predators that compete for fish with artisan fishermen, damage property and can attack humans. 'The Nile crocodile', writes McGregor, 'was, and is, widely disliked and much feared' (2005: 355). Local indigenous cultures thus view the crocodile with a mixture of reverence and fear. It appears in local mythology as a symbol of power and chiefship, but is also feared for its threat to human life and property, and as an element in witchcraft, reflecting the belief that witches travel as crocodiles. The hostile attitude of local communities to the crocodile was shared by the colonial authorities, who represented the crocodile as a pest and supported the growth of crocodile hunting as a way of controlling the species. The subsequent reduction in crocodile numbers was dramatic enough to put it on the list of endangered species, producing a new discourse of the Nile crocodile as an animal in need of protection. Conservation measures have accordingly set up protected zones for the crocodile and restricted hunting. In practice, this means that space has been given to the crocodiles at the expense of the community of artisan fishermen. Local communities have unsurprisingly tended to oppose such measures, which they perceive as an external imposition. Local indigenous discourses, meanwhile, continue to be marginalized in the management of the crocodile population, with McGregor noting that the re-evaluation of the crocodile 'did not draw on the opinion of local African communities, whose persistently hostile attitude towards the animal was deemed obstructive (to the extent it was considered at all)' (2005: 366).

Conflicts such as that over the conservation of the Lake Kariba crocodiles, or around the reintroduction of wolves in parts of southern Europe (Buller, 2004), provide fairly dramatic examples of the clash of different discourses of nature. However, even more mundane arenas of human–nonhuman interaction, such as between farmers and livestock, can be more complex and contingent than the taxonomy outlined at the start of this section suggests. Wilkie (2005) demonstrates this with a study of the feelings and attitudes of farmers in Scotland towards livestock. She proposes that these relations fall into one of four categories, dependent in part on the character of farming practice. Commercial family farming, she suggests, is characterized by 'concerned detachment', in which 'workers regard livestock as sentiment commodities' (p. 228). As farmworkers become detached from the everyday responsibilities of feeding and tending livestock, though, in industrialized agriculture, concerned detachment gives way to 'detached detachment', with animals perceived only as commodities. In contrast, Wilkie argues that hobby farming is typified by 'concerned attachment', 'where livestock are decommodified but can be recommodified at any time' (*Ibid*), whilst a privileged minority can afford to treat livestock as pets, such that they become totally decommodified in a relationship of 'attached attachment'.

A recurring problem in all these attempts to negotiate the co-existence of humans and non-humans in rural space is the apparent denial of the agency of animals and plants (see Jones, 2003). Nature does not easily conform to the assumptions of human attempts at control. Jones observes that 'cows, pigs, horses, poultry and sheep often have to be coerced by violent means into rural production networks' (p. 296), whilst wildlife may fail to behave as conservation strategies anticipate.

In an attempt to recognize the agency of the non-human in rural space, Cloke and Jones (2001, 2002) employ the concept of 'dwelling' that describes 'the rich intimate ongoing togetherness of beings and things which make up landscapes and places, and which bind together nature and culture over time' (Cloke and Jones, 2001: 651). Derived originally from the work of Martin Heidegger and modified by Ingold (1993, 1995), dwelling theory focuses on ways of 'being-in-the-world'. Ingold's particular contribution has been to connect dwelling to the production and reproduction of landscape and place, which he argues are created by contextualized, lived practices. As Cloke and Jones explain:

> Ingold tells us that landscape is the world as known to those whose practical activities take them through its manifold sites and who journey down its manifold paths. Landscape is where the past and future are copresent with the present – through processes of memory and imagination. Past, present and future are continuously reprocessed while the materiality of landscape is work by, and marks, this process.
>
> (Cloke and Jones, 2001: 652)

Critically, Ingold does not restrict the creation of landscape to human agency, but also acknowledges the actions of non-humans. In this way, Cloke and Jones suggest, 'dwelling allows actors other than humans a creative role' (2001: 653). Trees, for example, are creative agents in producing the countryside as both a living and a working space. As Cloke and Jones (2002) document, 'trees can construct places and vice versa' (p. 86).

Trees can be important landmarks for local identity, they can provide shelter and act as meeting places, or hiding places from the everyday surveillance of rural communities, and they can act as symbolic links to past events. They can also have an economic function, as illustrated in the example of an apple orchard in Somerset, south-west England. The orchard is a demarcated site of commercial production, in which the trees have been planted by people, and are enrolled by human actors in the production of fruit, which is harvested for sale as eating apples and for cider-making. Yet, Cloke and Jones (2002) argue that 'the trees bring to this process the unique creativity of being able to produce fruit in the first place' (p. 129). The orchard is hence a place that is co-constructed by human and non-human actors, whose entwined practices of dwelling have created landscapes with economic, social and cultural meaning.

Furthermore, Cloke and Jones (2002) suggest that a critical application of dwelling theory problematizes questions of the spatial boundedness of rural space. Recognizing the co-construction of rural landscapes by both human and non-human actors, and the necessary messiness of these 'thrown-together' relations, dissolves the spatial ordering of rural space between different types of nature discussed at the start of this section. However, dwelling has also frequently been equated with spatial boundedness in terms of 'rootedness' in place and the production of local distinctiveness, echoing ideas contained in some discourses of rural community, including that of new agrarianism. Yet, Cloke and Jones argue that such representations exhibit a potentially sinister rural romanticism that can reinforce the exclusionary character of rural community. Rather, they propose that 'if dwelling is to be a serviceable concept for contemporary landscapes, it needs to shed this reliance on local boundaries and instead reflect a view of space and place which is dynamic, overlapping and interpenetrating' (p. 138). Thus, they suggest that the Somerset apple orchard is more than a bounded space, but is engaged in networks that include the use of orchards as icons of regional identity, the specific marketing of 'English' apples, the material mobility of the fruit through processes of processing, sale and consumption, and the movement of the orchard owners, workers and other visitors, who 'all live spatially complex lives which take them through all manner of spaces both practically and imaginatively'.

CONCLUSION

The countryside is a lived space as much as it is a space defined by economic processes of production or consumption, as discussed in Chapters 3 and 4. It is also a space that is associated with certain ways of living. In particular, the notion of community has traditionally been invoked to represent the perceived close-knit interaction and solidarity of rural life. Meanings of rural community are discursively constructed in opposition to those of urban society, and, at an individual level, draw on ideas of the rootedness of rural people in place and in particular landscapes. Yet, communities are not only imagined, but are given material form through practices that are performed in certain spaces and structures. Collectively, discursive and material dimensions of rural community establish a framework for belonging to a community, but belonging comes with conditions and boundaries, such that rural communities can also be experienced as places of exclusion, marginalization and regulation.

Moreover, traditional constructs of rural community have increasingly been challenged by the effects of social and economic restructuring. The notion of a rural community as a settled, bounded collective has been disrupted by the enhanced mobility of rural populations. The out-migration of people from rural areas, especially young people, has depleted the critical mass of residents present to participate in community practices and support community institutions; whilst in other contexts, the arrival of new in-migrants to rural communities has challenged notions of belonging and generated conflicts over the meaning and practice of community. Additionally, rural space is increasingly home to many transient populations, including second home owners and migrant workers among others, who have multiple sites of belonging. Indeed, the mobility of rural people through and out of rural space has itself contributed to a stretching of community and the emergence of 'translocal villages'.

At the same time, attention must also be directed towards the coexistence of humans and non-humans in rural space. Being able to live with nature is a key element in many traditional discourses of rural life, identified as an attribute that marks out true rural people, and is founded on a taxonomy of non-human life that assigns different animals and plants with different functions and allocates them to different spaces within a spatially ordered countryside. However, the maintenance of these ordered relations has itself been challenged by the effects of social and cultural change, and particularly by the rise of alternative discourses and practices in which the conservation of non-human life is put equal with, or even before, the traditional interests and practices of rural communities. Renegotiating these relations requires a perspective in which the active role of both humans and non-humans can be accommodated, as arguably provided by the notion of dwelling. This in turn leads us back to the question about how being in rural space is performed, and as such to the performance of the rural, discussed in the next chapter.

FURTHER READING

Ruth Liepins (who now writes as Ruth Panelli), outlines her model for researching rural community in two papers published in the *Journal of Rural Studies* in 2000. The first reviews different approaches to community employed in rural studies and develops the new conceptualization, whilst the second paper applies this in empirical studies of rural communities in Australia and New Zealand. The practice of belonging in rural communities, and the implications of this for questions of inclusion and exclusion, is explored in work undertaken in England by Sarah Neal and Sue Walters, published in articles in *Geoforum* (2007) and *Sociology* (2008). Michael Bell's ethnographic study of a village in southern England, *Childerley* (1994), is an excellent examination of the dynamics of contemporary rural life and in particular the impact of in-migration. Sonya Salamon's book *Newcomers to Old Towns* (2003), similarly explores in-migration and community change in the Mid-West of the United States. A good summary of the literature on counterurbanization is provided by Clare Mitchell in the *Journal of Rural Studies* (2004), whilst Aileen Stockdale summarizes the dynamics, effects and experiences of out-migration from rural communities in *Sociologia Ruralis* (2004). The concept of the 'translocal village' is developed by Selvaraj Velayutham and Amanda Wise in a paper in *Global Networks* (2005) that looks at migrants from rural southern India in Singapore. The work of Paul Cloke and Owain Jones on dwelling and the creative agency of trees in the rural landscape is published in *Tree Cultures* (2002), with the main argument and case study of the West Bradley orchard in Somerset also presented in a paper in *Environment and Planning A* (2001).

7

PERFORMING THE RURAL

INTRODUCTION

The preceding chapters have focused on the discursive and the material construction of the 'rural'. It has been argued that the 'rural' exists primarily as an imagined category, a social construct that is brought into being through the production and reproduction of discourses that make claims about what the rural is, where it is, and how it should be managed. Yet, as the preceding chapters also demonstrate, the rural is also grounded and given material form. The materiality of the rural is both a product and a reflection of prevailing discourses of rurality, and can also serve to constrain and inform the discursive construction of the rural. The family farm, for instance, is a key motif in many imaginings of the rural, but it is also a distinctive material entity. As a material entity, the family farm has impacts on the local rural economy, rural environment and rural society, which in turn influence the discursive reproduction of the rural.

The missing link between the discursive and the material realms is the performance of particular actions by the diverse array of actors in rural space, both human and non-human, both 'local' and outsider. As Edensor (2006) suggests, 'it is through the relationship between the array of characters playing out particular roles, and the spaces in which they perform, that ruralities are routinely produced' (p. 484). However, performance is more than just a medium of translation – the enactment of both routine and staged performances can come to be seen as critical to the creation and articulation of rural identities. As examples discussed later in this chapter will illustrate, the routine performance of farmwork and the ritualized performance of hunting can both constitute the essence of the rural way of life for participants, such that individuals who have been in some way prevented from enacting these performances can feel that their rural identity has been compromised. In this way, performance is central to the everyday lives of the rural that constitute one dimension in Halfacree's three-fold model of rural space (see Chapter 1).

Shifting perspective to the performance of the rural can also bring to light dimensions of the rural, and power relations within the rural, that may be overlooked in other approaches. First, by placing an emphasis on the actions of the body, the concept of performance can permit investigation of the 'more-than-representational' geographies of the countryside (Carolan, 2008). Whereas the discussion so far in this book has tended to engage with various ways of representing the rural (as articulated through texts such as policy documents, newspaper reports, photographs and artwork, advertising brochures, individual stories, maps, diagrams, statistics, academic writing and so on), a study of the performance of the rural needs to be attuned to ways of knowing and living in the rural that cannot be easily represented, but which are about emotion, sense, instinct, intuition, habit and action.

Second, examining the performance of the rural can also expose details of the gendered construction and practice of rural life. For example, practices associated with hunting or drinking may be, in different ways, seen as part of the performance of masculinity in rural communities, as discussed further below. These associations help to reproduce a 'macho' rendering of masculinity, which serves in turn to underscore the patriarchal nature of rural society. Similarly, the gendered division of labour on the 'family farm' relates to the embodied performance of farm work, and the ways in which different tasks are deemed to be suitable for male and female bodies.

This chapter explores the performance of the rural from a number of different angles. After a short discussion that expands further on the conceptual foundations of the approach, the first perspective focuses on the performance of rural community. Building on the framework adopted in Chapter 6, this section examines the practices of rural community-making, looking first at the bar or public house as a site in which community is enacted through routinized yet stylized performances, and second at staged community events such as agricultural shows. The next section explores the performance of farming, repositioning the farm as an emotionally charged 'lifescape' in which certain practices and performances are routinely enacted, and gender distinctions reproduced and reinforced. The third perspective concentrates on hunting and its association with rural identity, examining the ties between the rituals of fox-hunting and the performance of rural community in England, and considering the significance of hunting to the construction of rural masculinity in Norway. The final section turns attention to the (staged) performance of perceived rural practices and customs by in-migrants, tourists and urban residents as a form of cultural consumption of the rural.

EMBODIED PERFORMANCES OF RURAL LIFE

The performance of the rural can take many different forms, with varying degrees of staging and scripting. Edensor (2006), in a seminal article sketching out a framework for research on performing rurality, notes that 'different rural performances are enacted on different stages by different actors: at village greens, farm-life centres, heritage attractions, grouse moors, mountains, long-distance footpaths and farmyards, and in rural spaces identified as "wilderness"' (p. 484). As such, Edensor includes within his definition of rural performances highly staged and scripted productions such as the representations of rurality enacted by many heritage attractions (Plate 7.1); more improvised yet still choreographed events of rural community, such as local shows and fetes; embodied experiences staged for tourists, from wine-tasting to whitewater rafting (see also Chapter 4); and routine, everyday practices of rural life. Yet, he contends, notions of rural performance and staging can be applied in each of these situations, and analysed with regard to the kinds of scripts and roles, and forms of stage management, choreography, improvisation and reflexivity, that contribute to the performance.

In several of the cases discussed by Edensor, the 'script' is provided by the particular representation of the rural that is conveyed through the performance. Thus, guides at tourist attractions and folk dancers at local festivals are enacting a particular representation of the countryside, often supported by a written script, or by visual and aural props such as costumes, pictures and music, that equally 'represent' rurality in a particular way.

Plate 7.1 Staged performances of rural heritage, Stundars Historical Village, Finland (Photo: author)

Indeed, some rural performances are explicitly about the performance of music or poetry that represents the rural, such that the performance exists as a way of communicating or enacting the representation contained in the text (see Box 7.1).

However, as performances invariably involve bodily action they cannot be reduced to text or representation alone. To perform rurality is to engage with rural space in ways that are sensuous, emotional and instinctive. For example, Carolan (2008) refers to a farmer interviewed in rural Iowa, who suggested that 'instead of knowing the countryside from a car on the road looking out at a field, I know it from my tractor in the field looking out at the road' (quoted by Carolan, 2008: 413). This was not

Box 7.1 PERFORMING RURALITY THROUGH MUSIC

Music has long provided an important vehicle for performing narratives about rural life, in many cases rooted in the landscapes and lived experiences of specific rural regions. Several musical genres have strong rural associations, including country in the United States and Australia, blues in the Mississippi delta, and folk music across the world. Folk music developed through the endogenous cultures of rural localities and has variously reflected the customs and practices of everyday rural life, commemorated key historical moments, and documented the challenges presented by rural change. In so doing, however, traditional folk music came to represent a nostalgic idyllized version of rural life, such that the collection and editing of 'traditional' music in the 'first English folk revival' of the late nineteenth and early twentieth centuries, for example, 'had less to do with preserving the past (which in the sense of rural community life was ever more distant) than with invoking a particular historic image' (Connell and Gibson, 2003: 38).

A number of present-day folk artists have resurrected the documentary and political tradition of folk music through song-writing that seeks to represent life in the contemporary countryside. The music of Show of Hands, a folk band from south-west England, is one such example. As Yarwood and Charlton (2009) describe, the band's music is strongly embedded in the geography of the English West Country, with frequent name-checking of towns, roads and rivers, as well as the use of actual landscapes as settings for the stories told through lyrics. In some cases, Yarwood and Charlton observe, the lyrics reinforce well-known place associations (such as mining in Cornwall, fishing on the North Devon coast), but through stories of motorway travel and small-town boredom and violence they also introduce a more contemporary edge. Indeed, with contemporary references to roads and motorways, in-migrants and drug-pushing, as well as historical references to seafaring, migration and the slave trade, 'their songs develop a relational sense of place that has meaning both in and outside the locality', avoiding presenting the south west as a closed region, but instead acting 'as part of a trans-border network that links and performs different spatial identities' (Yarwood and Charlton, 2009: 201).

Yarwood and Charlton also note that the songs of Show of Hands frequently connect distinct places, issues and times through lyrical associations. They cite as examples songs including, '"The Flood" (the flooding in Southern England and the travails of illegal immigrants); "Cold Frontier" (the memory of a Roman soldier on the fringes of the Empire with integration in today's EU); "Cousin Jack" (the poverty driven migration of the past/present pressures on Cornwall); [and] "Poppy Day" (a London drug dealer working smaller towns in the M4 corridor and a friend fighting in Afghanistan)' (*Ibid*: 202).

The countryside represented by Show of Hands is multi-faceted and polyvocal. Yarwood and Charlton list the characters appearing in their music as including the rural poor and excluded, bored teenagers, members of religious sects, petty criminals, gamblers, drug dealers, newcomers, poachers, sheep stealers, the voluntary emergency services, emigrants and military personnel. Different songs can relate different sides of shared experiences of contemporary rural life. For example, Yarwood and Charlton observe that two views of rural gentrification are conveyed in the songs 'Red Diesel' and 'Raining Again': '"Red Diesel" gives voice to a minor criminal who complains that "there ought to be a law for keeping out the yuppies and the grockles [tourists] (and the French)". By contrast, "Raining Again" deals with the semi-autobiographical experiences, hopes and aspirations of a couple moving to Dorset from London and captures some of the problems of achieving a bucolic vision of rural life' (*Ibid*: 203).

Accordingly, in as far as there is a political thrust to the band's music, it is against discourses that close down the definition of rural life. The song 'Country Life', for example, is noted for its critique of the Countryside Alliance and their mobilization of rural protests around discourse of the British countryside as defined by fox-hunting and farming. Rather, 'Country Life' 'blatantly disrupts and challenges established and establishment views of rurality

continued

(especially with the lines about "Landed gentry, country snobs" and second home owners)' (Yarwood and Charlton, 2009: 203). Yet, Yarwood and Charlton argue that in presenting this critique, 'Country Life' nonetheless formulated a representation of a discontented countryside, albeit with different emphases, and that as such, the song 'far from disrupting a rural idyll, reinforced the view of the countryside in crisis and, consequently, has been used by groups seeking to promote this view of UK rurality' (*Ibid*).

Further reading: Yarwood and Charlton (2009).

just a matter of perspective, as the farmer acknowledged with Carolan's prompting:

> Well, you get to know first hand the contours of the land as well as soil type – you know, what spots are wet, which are sandy, stuff like that ... When you've been doing this as long as I have you can tell a lot about the ground by how the tractor handles. It's almost like the tractor is part of me.
>
> (Iowa farmer, quoted by Carolan, 2008: 413)

In other words, it is the performance of farm work, the practice of driving a tractor day after day, that connects the farmer to the countryside in embodied ways that go beyond representations of the rural through texts and images. Such embodied performances, and ways of 'thinking as bodies' (Carolan, 2008), are as important as representational forms in the enactment of practices that constitute rural identity and rural being.

PERFORMING THE RURAL COMMUNITY

In her conceptualization of rural community (discussed in Chapter 6), Liepins (2000a) identifies practices as one of the key domains of community-making, alongside and inter-connected with community meanings (discourses) with spaces and structures (materiality). Community practices are informed by community meanings, and in turn reinforce these meanings through their enactment. They are similarly enacted within the spaces and structures of the community, can serve to animate these spaces and structures, but at the same time are constrained by the architecture of community spaces and structures. It is in this sense that we can think about the performance of rural community, as a necessary component in the constitution of community, but also as something that is intrinsically entwined with both the discursive representation of community and its material form.

Community practices, or performances, can be enacted in many different ways, as demonstrated in Liepins's (2000b) case studies of communities in Australia and New Zealand. At the most everyday level, community is performed through conversations in shops, post offices and at the school gate, the exchange of gossip and the smiles, nods and waves that constitute the unspoken acknowledgement of recognition between neighbours. At a more instinctive and more intuitive level, community is also performed through embodied practices of behaviour, dress, ways of speaking, movement, and a sensed knowledge of the community, its landscape and its people. However, community is also practised through more overtly staged and choreographed performances, such as annual events and festivals, village shows and fetes, cultural traditions and so on.

These practices help to construct community by bringing people together and circulating shared knowledge and experiences, providing the 'glue' that builds social capital. Yet, they are also mechanisms for articulating community identity, for both internal and external consumption, and, in particular, for aligning the identity of the community as being distinctively rural. Furthermore, by reinforcing norms of behaviour within the community, the performance of community practices can both cohere and regulate rural community life. Both of these elements can be observed in the examples discussed below of everyday interaction in the village 'pub', and of rural shows.

The pub is the hub

'The pub is the hub' was the slogan of a campaign launched by the Prince of Wales in Britain at the start of the twenty-first century, which sought to save country pubs from closure, arguing that they formed the heart of rural community life. In small communities with limited facilities, the local public house, 'pub' or bar, has a function that goes beyond the sale and consumption of alcohol. It becomes a multi-purpose social space for the community, the meeting venue for local clubs and societies, the organizational point for sports teams, and, in some cases, the host for postal facilities, medical sessions and even religious services. The pub hence serves as a key space in which community is performed, identity constructed and articulated, and social norms reproduced.

Moreover, the village pub is additionally a powerful symbol of rurality, such that the performance of community and the performance of rurality are blurred in the practices enacted in the pub. As Campbell (2000) observes:

> Rarely has any social site been mythologized to the same extent as the rural pub. From the idyllic haven of the English rural pub, to the last-chance saloon of the American West, to the crocodile-wrestling mateship of a corrugated iron shed somewhere in the Australian outback, rural drinking sites have been ascribed by both the popular imagination and academic analysis with pronounced mythic qualities. Such lay mythologies place the rural pub squarely within what rural sociologists term the 'rural idyll'.
>
> (Campbell, 2000: 562)

The myth is articulated in part through the material structures of the pub – its external appearance and interior design, the presence of an open fireplace, the pictures on the wall, the availability of regional beers and ciders, the food that is served, the clothes that are worn by regular customers and so on. It is also articulated, though, in the performances enacted within the pub, which are often presumed to include the social mixing of different residents of the community; the exchange of gossip and news; discussion of 'rural' topics such as farming, hunting and shooting; and the playing of traditional games such as dominoes or skittles. In this mythic representation, the English village pub also hosts hunt meets, or is the post-match venue for shooting clubs or cricket teams, becoming woven into the practice of these community activities as part of an extended social field of rural life. Furthermore, the cultural performance of some country pubs has been traditionally suspected of harbouring a flexible attitude towards certain legal boundaries, such as after-hours drinking, under-age drinking, gambling and drink-driving, reflecting Neal and Walters's (2008) observation that whilst rural communities may be places of internal regulation, they can also be sites of condoned deviance with respect to particular aspects of national laws and conventions as spaces that are beyond the easy surveillance of the state.

It is possible in countries such as Britain to find many rural pubs that exhibit one or more of the above characteristics, but Maye *et al.* (2005) rightly warn against 'the uncritical acceptance of the pub as a functioning structure of rural society' (p. 834). They argue that rural pubs are more diverse in character than the mythic representation suggests, existing on a continuum between 'folk culture' and 'popular culture'. In particular, they contend that many British country pubs have been shifted towards the popular culture pole through renovations and reinventions that have turned them into spaces of cultural commodification. Such 'standardized' and 'reconstructed' pubs are settings for staged performances of rurality, most commonly by serving locally sourced food, but also through live folk music or special events that purport to celebrate aspects of country life. In this way they attract a clientele that extends beyond the immediate community, including tourists, and may indeed be abandoned by local residents who no longer see them as a 'drinking pub'. Ironically, therefore, the pubs that might superficially most closely fit the mythic image of the country pub, and which might have the strongest ties to the local economy through supply chains, may also have the weakest degree of embeddedness in the local community and may be limited as spaces of community performance, although they continue to facilitate staged performances of imagined rurality (Maye *et al.*, 2005).

Pubs can also act as places in which social distinctions within rural communities are produced and reproduced. First, rural pubs are commonly highly gendered spaces, with the clientele and culture heavily male dominated. Not only might informal convention restrict the presence of women as customers (with their partners) to certain nights of the week (Heley, 2008), but the prevailing practices and performances at other times may frequently be encoded with a macho masculinity. Campbell (2000), for example, in a study of two village pubs in South Island, New Zealand, observed that 'the public performance of drinking actually involved intensely competitive interaction that resulted in distinct hierarchies among the men present' (p. 569). This included 'continual conversational cockfighting, during which other drinkers scrutinized men's performance' (*Ibid*), with jesting over drinking capacity and sexual prowess, but also references to local and rural knowledge and practices, with 'occupations outside the acceptable range of manual laboring, agriculturally related activities were also feminized derisively' (Campbell, 2000: 576). Accordingly, Campbell argues that 'male pub drinking practices have not persisted as a nostalgic memorial to a simpler life: they persist because they are a site of male power and legitimacy in rural community life ... rural pubs can actually operate as a key site where hegemonic forms of masculinity are constructed, reproduced and successfully defended' (p. 563).

Second, pubs can also be sites of class distinction within communities, or of differentiation between 'local' residents and incomers. Heley (2008), in an ethnographic study of a country pub in southern England, for example, observes the ostentatious display of wealth by moneyed in-migrants, who loudly perform acts of buying large, expensive rounds of drinks, and also notes the inscribing of social distinction into the geography of the pub, with one corner, 'the office', implicitly reserved for the established village elders. In communities that have been able to sustain two or more pubs, the social differentiation of customers can mean that the pubs become 'places where separate group identities are established and social boundaries maintained and legitimized' (Maye *et al.*, 2005: 834). A divide of this nature was observed by Bell (1994) between the two pubs in his study village in southern England. One pub, The Fox was favoured by middle-class villagers and was the scene of elaborate performances of conviviality:

> In the public bar of the Fox, people tend not to stand or sit oppositionally. Most of the men, particularly on a big night, will huddle around the bar in a great mass. In this huddle with its constant surging motion, it is faces that one directs attention to, rising and almost floating above a dark, lower zone of bodies, barely distinguishable from each other. Formal scrutiny of what would be called the 'body-face' is not possible. In the midst of this mass, one feels almost palpably part of a single group entity.
>
> (Bell, 1994: 58–59)

In contrast, customers in the second pub, The Horse and Hounds, favoured by the village's working-class residents, would 'sit directly opposite each other in small separate groups, backs facing backs at other tables' (p. 59). Yet, the less explicit performance of sociability in fact reflected a group more at ease in its identity, where performances of community took different forms – the practice of buying rounds was more commonplace in The Horse and Hounds than in The Fox, whilst the social space of the former also extended into a pub soccer team.

Rural communities on show

A second arena in which rural community is performed is in the plethora of annual events that are found across rural regions; including village fetes and festivals, carnivals, agricultural shows, state and county fairs and various local traditions, customs and rituals. These events serve a number of different functions. In many cases they are staged performances aimed primarily at tourists and visitors, and thus contribute to the local economy as a commodification of rural culture (see Chapter 4). In order to stand out as distinctive, rural events frequently purport to be performances of long-held local traditions, reflecting an authentically 'rural' culture, yet, they are commonly 'invented traditions', established in recent decades (often as part of 'bottom-up' rural development initiatives, see Chapter 5), and drawing only loosely on historic practice. The selective resurrection and re-presentation of local rural heritage in such invented traditions, in turn contributes to the articulation of community meaning, for both internal and external consumption. Nevertheless, the participation of local residents in such events can also practically help to reinforce community coherence, acting as a shared communal endeavour, or as a literal 'meeting-place' for community members.

Each of these three elements can be observed in the case of the 'Lights of Lobethal Festival', discussed by Winchester and Rofe (2005). The festival is a 17-day annual event preceding Christmas, in which residents of the small town of Lobethal, South Australia, decorate their homes and businesses with Christmas lights, as well as staging various Christmas-themed activities. The festival is claimed to be an expression of the town's Germanic Lutheran heritage, yet it is an invented tradition, initiated in 1947 and expanded in the 1990s following the closure of the town's main employer, a woollen mill. Today, the festival makes a major contribution to the local economy, attracting over 250,000 visitors annually, but as Winchester and Rofe (2005) note, it is also perceived to be critical in community-making, with one resident commenting that it 'brings us all together. It doesn't matter which area of the community you belong to, you've got the common denominator of Christmas time when we all join together' (quoted in Winchester and Rofe, 2005: 274).

The role of annual events in reinforcing community cohesion is especially important in areas where the rural population is dispersed in isolated farms and homesteads, and where shows and festivals form an excuse for community gatherings. Historically, this was a key function of agricultural shows, which along with regular markets provided an opportunity for farm households to meet up and for knowledge to be disseminated within the farming community. However, as Lewis Holloway (2004) documents, agricultural shows have evolved with rural change to become more outward-looking, providing agricultural societies with 'a context where they have the ability to stage-manage the presentation of agriculture to large numbers of non-farming visitors' (p. 320). Agricultural shows have thus become spectacles, in which the parading of show livestock and displays of vintage and modern farm machinery are part of a staged performance of farm life. At the same time, side stalls and demonstrations of rural crafts and customs present an opportunity for visitors to enact a rural experience that is far more varied in scope, as Edensor (2006) observes at the Cheshire Show in England. Noting that 'the proportion of farm machinery and livestock has dwindled' (p. 490), he contends that 'there was no single production but a multiply staged countryside in which hundreds of stages (stalls, displays, show arenas, demonstrations, play areas and information centres) vied with each other for the attention of the large crowds who wandered around the extensive site' (*Ibid*). Among the competing attractions were 'mounds of bread, jams and honey, cheeses, cakes and biscuits ... loaded on to rustic tables' (*Ibid*), craft products ranging from walking sticks and shooting sticks to wood-carvings and pictures of rural scenes, expositions of country skills, and exhibits of vintage tractors and rare breeds that required specialist rural knowledge to fully appreciate.

In this way, agricultural shows can be engaged with on different levels. For most visitors a show is an occasion to perform a simulated rural lifestyle for a day or afternoon. For an inner group, however, the agricultural show is part of the reproduction of their community, performed through active involvement in the showing and judging of livestock and produce. Entry to this inner community is determined by knowledge that permits an appreciation of quality livestock or good farm husbandry. Whilst Holloway (2004) suggests that shows are increasingly viewed as opportunities to educate the public about agriculture, the requisite knowledge cannot all be easily communicated, but extends into the realm of embodied and non-representational knowledge discussed by Carolan (2008), drawing on direct experience of working with livestock.

Rural events are hence simultaneously sites of inclusion and exclusion. They can contribute to community integration, but can also implicitly exclude groups who do not conform to community norms. Resurrected rural traditions, for example, can reproduce racial or gender stereotypes that can be offensive to groups both within and outside the community, leading in some cases to contestation and conflict (see Smith (1993) on the Peebles Beltane festival in Scotland, for example). At the same time, though, marginalized groups can use their own festivals and traditions to construct and perform alternative rural communities. The Appleby New Fair in northern England, for example, acts as an important economic and cultural event for the nomadic Gypsy-Traveller community in Britain (S. Holloway, 2004, 2005). Originally an occasion for sheep, cattle and horse trading that engaged the wider rural community, by the early twentieth century the fair had become primarily focused on horse-trading and dominated by Gypsy-Travellers. Between 5,000 and 10,000 Gypsy-Travellers attend the fair annually, both for trade and to socialize and meet with fellow members of their dispersed community. However, Gypsy-Travellers have repeatedly been the targets of discrimination in the modern British countryside, and the annual convergence of a large Gypsy-Traveller community on a small town of fewer than 3,000 permanent residents generates tensions. There have been attempts to ban the fair, and, as Holloway (2005) shows, many local residents portray the Gypsy-Travellers as different and as sources of disruption. The disjuncture between the form of rural community performed by the Gypsy-Travellers at the fair – mobile and disorderly – and the settled, property-based and ordered community performed by local residents and farmers sits at the heart of the tension, but some residents differentiated between 'true Gypsies', who they accept as part of rural tradition, and 'hangers-on', who are vilified as potential criminals. However, the survival of the fair arguably represents the acceptance in broader rural society that the countryside is multiple and that rural community can be performed in many different ways.

PERFORMING THE FARM

Emotional geographies of farm lifescapes

Geographical research on farming has conventionally focused on the economic and environmental dimensions of agriculture, constructing the farm primarily as an economic unit. Whilst rural sociology has had a stronger tradition of additionally recognizing the social dimensions of farming, and the significance of the farm as the site of complex labour and social relations, its approach has tended to emphasize structural formations, and the position of farming within broader meta-structures of class, gender and capitalism. It is therefore only with more recent calls for the application of a cultural perspective to the study of agriculture (Morris and Evans, 2004) that attention has begun to be paid to the farm as a discursive construct, and as a site of performance.

Although the discourse of modern capitalist agriculture encourages farmers to think of themselves as professional 'producers' or 'growers', and of their farms as 'businesses', the status of 'farmer' is not like any other occupational classification. To be a farmer is to conform to a particular way of life that has social, cultural and moral dimensions as well as an economic purpose. This identity is constructed and reproduced both through popular culture and through the lay discourses of farming communities, such that there is a social expectation on farmers to think and act in certain ways. Accordingly, the identity of being a farmer is also performed, enacted through working the land and through the farmer's relations with their family, their workforce and the wider community.

The farm is the site of the performance of farmer identity, and as most farmers live and work on their farm, they are connected to its landscape by a complex web of social, economic, cultural, moral and emotional interactions. Convery *et al.* (2005) refer to this as the *'lifescape'* of the farmer, borrowing the concept from anthropology where it was developed as a way of framing the social, cultural and economic engagements between people and landscape. As Convery *et al.* (2005) explain, when applied to farming, 'lifescape articulates the spatial, emotional and ethical dimensions of the relationship between landscape, livestock and farming community and elucidates the heterogeneity of agricultural emotional landscapes' (p. 101).

These complex relationships are evident in the oral histories of farming collected by Riley and Harvey (2007) in England. The farmers' narratives that they report are replete with an intimate knowledge of the farm landscape. Fields, hedges and artefacts such as an abandoned tractor, provided cues for stories and memories that were frequently entwined with biographical events, such as one farmer's recollection of mowing a field for hay on his niece's wedding day, between the ceremony and the reception. Furthermore, Riley and Harvey note that 'aspects of the past become embodied through the repetition of performative agricultural practice' (p. 403). Many of the stories that they quote from farmers involve memories of activity – ploughing, mowing, baling hay, taking cattle to a stream to drink, creating a water meadow – whilst comparisons were also drawn between past and present practices. As such, Riley and Harvey (2007) suggest that 'current practices are underpinned by a cumulative understanding that often stretches back over many generations' (p. 404).

The emotional ties between the farmer and the farm landscape are perhaps thrown into sharpest relief when external forces intervene to disrupt the performance of everyday, iterative farm practices, as during the 2001 epidemic of Foot and Mouth Disease in Britain. Efforts to control and eradicate the disease included the imposition of quarantine on farms (often voluntarily), and a cull of all livestock, including healthy animals, in areas proximate to confirmed outbreaks. These measures disrupted the everyday performances of farm life – farms where livestock had been culled were left empty of animals, with no need for daily practices of feeding, herding, milking and so on; whilst regular patterns of social interaction at markets and other community events were placed off-limits. Moreover, the cull of healthy livestock represented for many farmers the inversion of normal practice and a violation of the established lifescape, as Convery *et al.* (2005) describe:

> There was a clear breach of normal relations – whilst lambs are normally slaughtered, this is not when they are newborns, and so the rhythm and cycle of livestock farming relations was out of synchronization. The epidemic created fissures in the taken-for-granted lifescapes which transcended the loss of the material (i.e. livestock) to become also the loss of the self ([farmers'] perceptions of identity and meaning associated with this lifescape were called into question). Death was in the wrong place (the farm rather than the abattoir) but it was also at the wrong time (in relation to the farm calendar)

> and on the wrong scale (such large scale slaughter seldom occurs at the same time).
>
> (Convery *et al.*, 2005: 104)

Some farmers and rural residents affected by the Foot and Mouth Disease epidemic vented their emotions through poetry, much of which attempted to articulate the disrupted lifescape, with titles such as 'Where cattle grazed', or made reference to the changed performances of farm life, with animal husbandry replaced by the brutality of the cull and the everyday sounds of the farm silenced (Nerlich and Döring, 2005).

Gendered performances of farm life

A further peculiarity of agriculture as an economic sector is the welding together of the farm enterprise and the family household into a single entity in the concept of the 'family farm', as widely articulated in the global north (see also Box 7.2). This state of 'consubstantiality' (Gray, 2000), in which family and farm 'become united in common substance'

Box 7.2 GENDER AND FARM WORK IN THE GLOBAL SOUTH

Women farmers play a critical role in food production in the global south. As a response both to bereavement and to the migration of rural men to find paid work in cities (see Chapter 6), women in Africa, Asia and Latin America have increasingly assumed responsibility for both commercial and subsistence farming. The Food and Agriculture Organization of the United Nations (FAO) estimates that at least four in ten farmers in Africa are women, and that women are responsible for between 60 and 80 per cent of all African food production (Zaccaro, 2009).

Yet, the work of women farmers is frequently marginalized or rendered invisible in official discourses by cultural conventions that continue to construct farming as a male activity. These include perceptions about the bodily practices of farm work and the ability of women to perform these tasks. In the Amhara region of Ethiopia, for example, a 'farmer' is commonly defined as someone who can independently engage in the activities of sowing and ploughing. Women contribute around 40 per cent of agricultural labour in Ethiopia and are involved in sowing seed, weeding, harvesting, threshing, herding and tending livestock and milking, but because few women plough the land, they are not considered to be farmers (Frank, 1999). Ploughing is regarded in the local culture as an inappropriate activity for women and too physically demanding for the female body. However, as Frank observes, it is not too physically demanding for young boys, who frequently help with ploughing, and many women in the region say that they would like to plough, but are prevented from learning by men. As such, Frank concludes, 'perhaps women's inability to plow [sic] is based more on cultural perceptions than on actual physical inability' (p. 3).

The discursive marginalization of women farmers has serious structural and material impacts. Across the global south, women farmers own only a tiny fraction of land, and are constrained by less favourable sites and more restrictive land tenure and ownership conditions than male farmers. They are also less able to obtain credit or to access training, and have less of a voice in farming organizations or with government agencies. As Hart (1991) argues in research on Malaysia, these systemic biases reinforce the differential capacity of poor rural women and men to organize resistance to the interests of the state and large capitalist landowners, and to take advantage of agricultural reforms and opportunities (see also Angeles and Hill, 2009; Koczberski, 2007; Razavi, 2007).

Further reading: Hart (1991).

(*Ibid*: 345), is important in shaping the lifescape of the farm, as discussed above, and can be both a point of vulnerability and a source of resilience in confronting challenging conditions for agriculture (Johnsen, 2003). Fundamentally, the notion of the 'family farm' presents farming as a shared endeavour, in which all members of the family have a part to play, yet, as Brandth (2002) explains, the division of roles within the family farm is strictly gendered:

> Family farming is patriarchal; the male farmer is the head of the farm family and the family farm and makes the relevant decisions. He is the farm's public face, and he participates in agricultural organizations and forums. Family farming is based on the labour force of family members with the allocation of tasks being fundamentally gendered. Women are responsible for care and household tasks and this task allocation has been regarded as a 'natural' distribution of work on the basis of certain gender specific attributes.
>
> (Brandth, 2002: 184)

The gender division of labour on the farm is discursively reproduced in popular culture, in the farming media, and in the documents and rhetoric of agricultural organizations (Brandth, 2002; Liepins, 2000c), but is also performed through the everyday enactment of work on the farm. In this the male farmer will typically undertake tasks concerned with managing and working the land, handling livestock and operating machinery – in other words, the activities that are popularly understood as 'farming'. Farm women, by contrast, will typically look after the paperwork and farm accounts, carry out basic tasks of animal husbandry, run on-farm sidelines such as a farm shop or bed and breakfast accommodation, and maintain the farm household – in other words, activities that are not commonly understood to be 'farming', but which are nonetheless critical to the operation of the farm.

According to the discourse of the family farm, this gender division of labour reflected the greater physical strength of men, required for the physically demanding character of work on the land (Brandth, 2006; Saugeres, 2002a). However, as the physicality of farm work has been reduced by mechanization, male farmers have retained their primacy by constructing technical proficiency as a masculine attribute. Indeed, mechanization can be argued to have contributed to a defeminization of farm work, as farm women are no longer required to help with activities such as harvest, or to contribute bodily skills of dexterity in tasks such as tying bales of hay (Riley, 2009). The changing gender deployment of labour on the farm is captured in the reflection of one farm woman in the English Peak District, who recalled whilst preparing lunch in the farm kitchen that, 'I used to do a lot more of this work that the men are doing out there today ... it's all changed. My husband and sons do it all on tractors these days, so I'm in here and less involved with that' (Riley, 2009: 666). There is accordingly a gendered geography to farm work, with women confined to 'backstage' roles in farm kitchens and offices, whilst men perform on the 'frontstage' of the farm (*Ibid*).

Moreover, the inheritance of farms through the male line is evoked to construct farming as a masculine pursuit. As Saugeres (2002b) observes based on research with French family farmers, 'farmers are described as having a natural predilection for working the land. This predilection is said to have been acquired at birth and through inheritance giving the paysan his connection to the land' (p. 377). In contrast, in language that resonates with Carolan's (2008) sense of thinking as bodies, 'a woman is represented as not being able to farm on her own because she lacks an embodied knowledge of farming and an embodied connection to the land' (Saugeres, 2002b: 382).

Thus, the construction of farming as masculine is maintained both through discursive repetition and through repeated practice, as Saugeres (2002a) observes:

> Because farming is constructed as masculine, farmers' discourses and practices come to reinforce and legitimate the boundaries that maintain this space as masculine. These boundaries are maintained through a discourse emphasizing physical strength and a natural aptitude for technology as the two main qualities which are essential to being a farmer. As it is only men who are supposed to have these attributes and these are naturalized according to biology, this situation appears as part of an inevitable natural order and beyond the scope of change.
>
> (Saugeres, 2002a: 156)

Yet, because the gender division of farm labour needs to be enacted through practice, it can be contested and contradicted through performance. Both Saugeres (2002a) and Silvasti (2003) recount examples of women who have taken on traditionally masculine roles within farming. Such is the power of the family farm discourse, however, that the gendered division of labour is commonly treated as natural and inevitable by farm women as well as by male farmers (Brandth, 2002), and even women sole farmers may be inclined to judge their success or failure in terms of gender. Silvasti (2003), for example, cites one female farmer in Finland, Anne, who gave up her attempt to continue the family farm, stating that the 'job is too heavy for me' and that 'this kind of work is not suitable for a woman' (quoted in Silvasti, 2003: 161). As Silvasti comments, 'it is interesting that in the moment of defeat, Anne refers exactly to the deficiency of her own body. The economic problems were severe, adopting a social position as *isäntä* [a Finnish peasant farmer] as well as constructing her own identity as a farmer were both difficult, but finally she was disheartened by her own body – young, beautiful and deficient' (Silvasti, 2003: 161).

There are, however, some signs that the gender construction of farm work is shifting. Not only are numbers of independent women farmers increasing, but younger male farmers, in particular, are also reassessing their identity in the context of technological innovations in agriculture. Australian farming men interviewed by Coldwell (2007), for example, emphasized keeping up with technology and business skills as well as toughness for physical labour in their description of themselves. Whilst these attributes were still presented as an expression of masculinity, Coldwell argues that the reflexivity that they involve is part of a new 'dialogic masculinity' in which farm work is not closed off as an exclusively male preserve. At the same time, though, high levels of depression and suicide among farmers have been linked to the self-doubt of farmers who feel that they are unable to match the expectations of the masculinist model of farming in an increasingly difficult economic environment (Price and Evans, 2009; Ramírez-Ferrero, 2005).

PERFORMING RURALITY THROUGH HUNTING

The hunting of wild animals has historically been an important part of rural cultures in both the global north and the global south. Animals have been hunted for food and for clothing; to control dangerous pests and predators that have threatened crops, livestock and even human life; and for sport and recreation. For rural peoples, hunting is often evoked as an expression of their connection with nature, of their negotiation of sharing rural space with non-humans (see Chapter 6). Hunting symbolizes human domination over nature, but it can also be presented as stewardship of nature, and as a practice within a natural order of predators and prey.

The performance and ritual of hunting, though, can additionally have symbolic significance for the construction of endogenous rural identities (see Box 7.3), for the maintenance of social order within rural communities, and as a rite of passage for rural youth, as the examples discussed in this section demonstrate.

Box 7.3 HUNTING AND INDIGENOUS RURAL CULTURES

Hunting is an important cultural activity for many indigenous peoples, for whom hunting may be not only a performance of traditional rituals and customs, but also a key component of their spirituality and their relationship with place. As Perreault (2001) argues, 'place identity' is particularly important to indigenous peoples because land and land-use is often central to their way of life and their understanding of themselves. Traditional hunting can be a way of performing place identity, but it can also be a point of conflict with external values and environmental, political and commercial interests.

For the Inuit of Arctic North America and Greenland, for example, the collection of 'country foods' from the land, sky and water by hunting, fishing and gathering is fundamental to their identity and their connection to their environment. Gombay (2005) explains that as Inuit hunt, fish or gather food, 'the material and immaterial worlds blend together, with layer upon layer of meaning and understanding. The getting of country foods is about understanding the land in which one lives. It is about building an awareness and knowledge of one's place in the natural world of living and nonliving beings' (p. 418). This means working with nature, knowing where and when to find plants or animals, and, according to Inuit spiritual beliefs, accepting the gift of the hunted animal as it presents itself to the hunter.

Such is the centrality of 'country foods' to Inuit culture that community members are inducted in hunting from childhood:

> From the time that they are babies they see seals shot, caribou butchered, and berries picked: they know what it is to eat raw

continued

> fish, still warm from the hook. As soon as they are able, they are encouraged to take part in these activities. Each point along the way in their development as a hunter is marked with pride.
>
> (Gombay, 2005: 419)

Crucially, tradition dictates that country foods are to be shared in the community, with sharing acting as 'part of the glue that binds community together' (*Ibid*: 420). However, these cultural customs have been challenged as the Inuit population has been settled in fixed communities and institutionalized in recent decades. A Hunter Support Programme (HSP) was set up by the Canadian government in the 1970s to pay hunters for collecting country foods, intended to support traditional culture. Food purchased through the HSP was stored and distributed free to community members, but as Gombay (2005) reports, tensions have developed in some communities over the sharing of HSP food – for example, in the distribution of fish or meat to non-Inuit residents.

Further reading: Gombay (2005).

In recent decades, however, growing awareness of the environmental impact of hunting in terms of depletion of species numbers, combined with an increased interest in 'animal rights', especially among urban populations, has put hunting under pressure. In the global south, conservation projects have attempted to discourage indigenous populations from hunting endangered species through education programmes and ecotourism initiatives (Gibson and Marks, 1995). In many parts of the global north, legislation has sought to regulate and restrict the practice of hunting as sport. Yet, in both contexts, anti-hunting strategies have frequently faltered in misunderstanding the cultural significance of hunting to rural communities, and the emphasis placed on hunting as a performance of rural identity.

Ritual, performance and hunting in rural England

The hunting of foxes and deer with hounds was a traditional and iconic pastime in the English countryside for over two hundred years, before it was outlawed on animal welfare grounds in 2005 (Woods, 2008b). The lengthy public debate preceding the legislation focused not only on questions of animal welfare and the contribution of hunting to the rural economy, but also on its symbolic significance in English rural culture. Hunting supporters argued that the proposed ban was an assault on rural identity, and the social and cultural importance of the sport in rural society is indicated by the continuation of its rituals and performances in modified, legal, form even after the introduction of the ban that specifically relates to dogs chasing and killing wild mammals.

The performance of hunting traditionally revolved around the chase, in which a fox or deer was pursued over several kilometres by hounds, followed by the 'hunting field' mounted on horseback, and various 'hunt followers' in cars, all-terrain vehicles or on foot. In legal hunting, a false scent rather than a live animal is pursued. As Marvin (2003) describes, the hunt is a scripted and choreographed performance with defined roles for participants (including the hounds, horses and the hunted fox or deer), rituals and conventions. The performance is supported by props, such as the Huntsman's horn that signals the start and end of the hunt, and the costumes of the mounted followers, which vary by season and office according to a strict protocol (Cox *et al.*, 1994).

A hunt is hence a spectacle, in which participants 'announce their presence in the countryside and draw attention to themselves both visibly and audibly' (Marvin, 2003: 49). Yet, it is a spectacle for internal rather than external consumption. Marvin (2003) contends that 'there is no distanced, unconnected, audience for fox-hunting, there are no spectators who are outside of the event itself and that there are no mere observers of it' (p. 52). All present at a hunt are participants in the performance of the hunt, and through it, it might be claimed, in the performance of rural community (Plate 7.2).

Hunting contributes to the performance of rural community by providing a meeting-place for rural residents, both at hunt meets and through fundraising social events – ranging from hunt balls to bingo and whist drives (Cox *et al.*, 1994; Milbourne, 2003). At the same time, belonging to the community defined by the hunt is established not only by attendance at these events, but also by familiarity with the distinctive rituals, customs and linguistic codes of hunting. As Cox *et al.* (1994) explain, 'like all ritual, this serves to confer a clear sense of identity and meaning for those involved and a correspondingly acute sense of exclusion for those not familiar with the mores of hunting' (p. 193).

Plate 7.2 Hunting and the performance of rural community, Exmoor, England (Photo: author)

Full participation requires not only an understanding of the language and customs, but also an appreciation of the skill of hunting and the work of the hounds. Hunt followers routinely engage in a running commentary and analysis of the hunt, drawing on a 'deep knowledge of hunting [that] is not simply directed towards informing a critical, distanced judgement of the performance of others but is rather an understanding directed towards an experience in and of the natural world for themselves. It allows them to be fully, immediately and actively present' (Marvin, 2003: 55). Possession of this knowledge is perceived as a distinctively 'rural' trait, an association that is reinforced by the movement of the hunt through the rural landscape. As Marvin explains:

> Hunt members are immensely committed to their Hunt, express a deep sense of belonging to a particular Hunt 'country'. It is a countryside of which they have a close and intimate knowledge because of their hunting experiences. It is not a countryside that they visit, it is a

> countryside that they are of and to which they belong ... The very landscapes and that which they contain are transformed into a performance space for the day. But that transformation, although new, fresh and full of immediate potential on each hunting day also has a long history that gives it a powerful depth and resonance for those who have regularly hunted across it.
>
> (Marvin, 2003: 50)

Marvin accordingly argues that the countryside is not merely a stage or setting for the performance of hunting, but rather 'is an active constituent of the event itself' (p. 51). Hunting and rurality are hence inextricably connected.

Hunting and masculinity

Outside Britain, hunting in the global north typically refers to the tracking and shooting of birds and mammals. In contrast to the performance of English fox-hunting, tracking and shooting is commonly practised by individuals or small groups. It is the hunter who personally tracks and kills the animal, and as such hunting is constructed as a test of physical and mental strength. In turn, these perceived attributes lead to a construction of hunting as a masculine activity, as can be illustrated by the example of elk hunting in Norway (Bye, 2003, 2009).

Hunting elk is a relatively recent tradition in Norway, but it has nonetheless assumed a position of cultural significance such that in the forested areas of Norway, 'elk hunting plays a central role in the shaping of the masculine rural identity' (Bye, 2003: 145). As a signifier of manhood, induction into hunting is a rite of passage for Norwegian men, enacted through steps of carrying a rifle, bringing down the first animal, and killing the first bull. As such, elk hunting 'is a symbol of and a ritual for important transitions in people's lives, exemplified by expressions such as "too young to go hunting", "elk confirmee", "hunter" and "retired hunter" (Bye, 2003: 145). Men in Norwegian rural communities are expected to participate in hunting: 'if a man living in a "hunting community" does not take part in hunting he may easily feel sidelined or "out of place", meaning he is not a "real man"' (Bye, 2009: 282). Masculinity and rural identity are hence entwined together in this discourse.

Elk hunting is a performance that adheres to certain conventions and rituals, requiring the hunter to act in particular ways. Correct costume and props are important, such that 'the perception of whether or not a man is an acceptable hunter or outdoor enthusiast is decided to some extent by his clothing' (Bye, 2009: 262–63), with 'true' rural hunters avoiding 'fancy and expensive Gore-Tex clothing' (*Ibid*). On the hunt itself, the hunter is required to exercise discipline and to demonstrate skill, as 'a particularly ugly wound (wounding an animal without bringing it down) or violating the party leader's instructions may involve a loss of prestige, and in a worst case scenario, exclusion from the hunting party and even from the local community' (Bye, 2003: 145).

The elk hunt provides a space in which bonds between rural men can be developed and community enacted. Comradeship is forged in the team-work of the hunt, but also through drinking in the evenings. In these ways, 'the male hunt is presented as a symbol of freedom because men can hunt at their own pace without having to show any special consideration for the women; they can talk freely about hunting and get drunk without worrying about others or what they think' (Bye, 2003: 150). Although the relative participation of women in hunting in Norway is increasing, the young men interviewed by Bye (2003) were ambivalent about women hunting. Some had been hunting with their partners, but they expressed concerns about the physical capabilities of women and at the loss of the exclusive male space.

This exclusive space is also defended from the intrusion of urban-based hunters, who participate in commercialized hunts as part of their own attempt to perform a perceived rural masculinity. It is, however, the failure of urban hunters to adequately perform as a hunter should – in the opinion of the rural hunters interviewed by Bye (2003) – that becomes the distinguishing characteristic between rural and urban men:

> Urban men are characterized as 'the other' in the sense that they represent the foreign substance in the rural environment. They are described as 'extravagant' and 'self-centred', and they are categorized as macho because they do not show any respect for the wild game and nature management. Moreover, the general opinion is that this group does not possess the necessary local knowledge to become good hunters. When the rural man is represented as the counterpart

> of the urban man, the rural masculinity is constituted and upgraded: the rural men become the epitome of the strapping and balanced man. They also represent realness and authenticity, because hunting for them is not only a hobby, but also a way of life.
>
> (Bye, 2003: 151)

(EX-)URBAN PERFORMANCES OF RURALITY

The preceding sections of this chapter have focused on various ways in which rurality is performed by people within the countryside, frequently emphasizing the significance of rituals or forms of knowledge that are exclusive in nature and which hence function to enact rural identity in exclusionary terms. However, the examination of performance can also shed light on the dynamics of engagement with rural culture and rural lifestyles by outsiders, notably urban migrants and consumers. Many urban engagements with the countryside involve the enactment of practices in which individuals attempt to 'perform' rural roles and characters. These include various types of rural tourism activities, ranging from the consumption of commodified heritage sites, through to embodied forms of adventure tourism and ecotourism, to 'working holidays' on farms and conservation projects (see Chapter 4). They also, though, include the efforts of in-migrants to rural areas to integrate with the local community and to participate in local events and activities.

Edensor (2006) proposes that whenever 'city folk' engage with the rural, as tourists or as in-migrants, they are confronted by the need to adapt to new performative conventions:

> Like all social performances, culturally specific ways of acting in rural theatres are organized around which clothes, styles of movement, modes of looking, photographing and recording, expressing delight, communicating meaning and sharing experiences are deemed to be appropriate in particular contexts. Initially, particular enactions are learnt so that the necessary competence is acquired, and the suitability of the performance is also likely to be subject to the disciplinary gaze of co-participants and onlookers. Through such socially constituted approaches to being and acting in rural contexts, urbanites gradually lose self-consciousness and self-monitoring as they become

> more grounded in 'common sense' and unreflexive assumptions about how and where to walk, how to 'appreciate' and comment upon beauty, how to climb, run, ski or relax.
>
> (Edensor, 2006: 486)

In making this transition, urbanites need to negotiate the differences between the commodified versions of rural lifestyle represented in popular discourses and the actual practices enacted by endogenous rural communities. The urban hunters derided by Bye's (2003) young rural men for wearing Gore-Tex clothing and buying fancy equipment fall into this trap – purchasing props that are marketed to urban consumers as essentials of rural life, but which are rarely used by rural residents themselves.

Heley (2010) similarly recounts the performances of affluent incomers to a village in England, who he describes as part of a 'new squirearchy'. This group aspires to the lifestyle of the old rural elite, and acts out perceived elements of this lifestyle through, for example, participation in hunting and shooting; occupying large houses and owning land; keeping horses and dogs or engaging in hobby farming; driving Range Rovers, or other four-wheel-drive vehicles; dressing in tweeds and Barbour jackets; and being a visible and vocal presence in the village pub and at community events. However, rather than integrating participants with the local community, these performances make them stand out. The selective enactment of certain lifestyle practices falls short of assimilation with the rural upper classes, particularly as the 'new squirearchy' performance is shorn of the moral and political obligations undertaken by the traditional squirearchy towards rural communities.

Furthermore, the mobility and mutability of many of the props and practices that are perceived by urban actors to constitute a rural lifestyle means that rural performances need not necessarily be tied to rural space. Both urban consumers enthralled with rural culture and rural migrants to cities have been responsible for translating certain simulations of rural life into urban spaces. These can include, for example, attempts at urban agriculture or the replication of 'rural nature' in urban and suburban gardens; the practice of traditional 'rural' crafts by city dwellers, including 'rural' cuisine, and the operation of shops and societies that support these activities; and celebration of traditional rural festivals by migrant urban communities; as well as the ostentatious use in urban contexts of consumer goods that imply the performance of a rural lifestyle, such as four-wheel-drive cars, branded clothing associated with rural connotations such as Barbour or Gore Tex, or household furnishings on a rural theme. However, performance cannot be entirely divorced from context, such that transposed into an urban setting, these props and practices facilitate not the performance of a rural lifestyle, but a hybrid performance in which urban and rural signifiers are mixed and blurred.

CONCLUSION

Rurality is not only constructed discursively and materially, it is also performed. The performances of people in the countryside, both residents and visitors, turn discursive representations into practice, and become ways of structuring life in the countryside. The routinized performance of everyday practices naturalizes discourses of rurality and the social relations contained therein. They cohere communities and organize the working of the land for agriculture. Some performances may be clearly staged events, explicitly enacting a particular representation of rurality for internal or external consumption. Yet, even routine everyday practices can be scripted and choreographed, in that participants perform prescribed roles and characters and act within established parameters.

The performance of rurality hence enacts rural identity and demarcates boundaries that delimit who is deemed to belong in a rural community, and who is not. Acceptance is not just about conforming to a discourse, it is also about acting in the correct way, using and understanding the right language, participating in rituals and traditions, and appreciating the way in which things are done. In this way, the performance of rurality includes embodied practices that are not representational, that concern ways of knowing and feeling through instinct and intuition and bodily perception. The practice of farming, the performance of hunting, and the enactment of a community's ties to a specific territorial place all involve a connection to the rural landscape and environment that is more than representational.

The act of performing rurality also enforces social distinctions within rural communities. The construction of certain practices as masculine and others as feminine entrenches gender relations in rural communities that restrict the ways in which it is acceptable for men and women to act, reproducing an essentially patriarchal society. Similarly, the capacity to enact certain performances within rural space may be dictated by class, or age, or ethnicity or religion. Breaching these conventions, by enacting practices and roles in contravention of these norms (or by failing to perform as expected) may attract suspicion from other community members, and may in effect be seen as a voluntary abdication of rural identity. Discourse and performance are therefore mutually constitutive: performance enacts discourse and the interpretation of performance is informed by discourse.

FURTHER READING

The best introduction to the concept of performing rurality is provided by Tim Edensor in his chapter the *Handbook of Rural Studies* (2006), which covers several of the themes discussed in this chapter. The more-than-representational dimensions of rural performance, and the idea of knowing the rural through the body, are discussed by Michael Carolan in *Sociologia Ruralis* (2008). For more on the connections between drinking culture, rural community and masculinity, see Hugh Campbell's paper in *Rural Sociology* (2000), whilst Lewis Holloway discusses the changing function of the agricultural show in the *Journal of Rural Studies* (2004). There is a fairly extensive literature on gender and farming, only some of which is referred to in this chapter. Berit Brandth's article in *Sociologia Ruralis* (2002) provides a good overview of the literature, whilst two papers by Lise Saugeres in the *Journal of Rural Studies* (2002b) and *Sociologia Ruralis* (2002a) form a good examination of the performance of masculinity on the family farm. Ian Convery, Cathy Bailey, Maggie Mort and Josephine Baxter, writing in the *Journal of Rural Studies* (2005), introduce the concept of 'lifescape' and describe the emotional impact of Foot and Mouth Disease in Britain in 2001. Garry Marvin provides a vivid description of fox-hunting as performance in his chapter in *Nature Performed*, edited by Szerszynski, Heim and Waterton (2003), whilst the rituals of stag-hunting and their part in the constitution of rural community are discussed by Graham Cox, Julia Hallett and Michael Winter in *Sociologia Ruralis* (1994). The significance of hunting to rural masculinity in Norway is explored by Linda Marie Bye in the *Journal of Rural Studies* (2009). For more on the performance of rurality through folk music, and particularly the example of Show of Hands, see the paper by Richard Yarwood and Clive Charlton in the *Journal of Rural Studies* (2009).

8

REGULATING THE RURAL

INTRODUCTION

There are many different actors involved in the production and reproduction of the rural: rural residents and in-migrants, farmers, corporations, tourists and visitors, the media, lobby groups, academic researchers, and a plethora of other social actors, as well as non-human actors including plants and animal, as discussed in previous chapters. This chapter focuses on the role of the state in the production, reproduction and regulation of the rural. The state's involvement is extensive and takes several different forms. The state is first of all active in *defining* the rural through official classifications of rural and urban areas that are subsequently employed in both the framing and the delivery of government policy, and through various non-governmental actors in lay interpretations of *where* is rural (see Chapter 2). The state is also involved in *describing* the rural, through the collation and analysis of statistics about the rural economy, population and environment, and through the production of maps and reports that portray and document the rural. These descriptions inform government policy, but they are mobile representations that can be translated and deployed in different ways by a variety of actors. For example, maps of the rural landscape initially produced by state cartographic agencies for military purposes, or to assist with the process of governing, are put to new use by tourists who utilize them to access the countryside for leisure and recreation.

The state is further active in the on-going *regulation* of the rural through the formulation and implementation of rural policies, with the capacity to reach into every aspect of rural life. Agricultural policy, for example, informs the conditions of production for agriculture, the viability of different types of farm organization and practice, and the financial health of individual farms; environmental policy influences engagement with nature; planning policy regulates how rural land is developed. More broadly, policies for health, education, transport, social welfare and so on determine the provision of public services in rural communities. Through these policy instruments the state is not only active in the production and reproduction of discourses of rurality, but also shapes the material construction of the rural. The state has the capacity to influence the appearance of the rural landscape, the structure of the rural economy, the pattern of rural settlement, the character of the rural population, the nature of rural education and health care, the presence of fauna and flora, the commodification of the countryside for tourism, and the standard of living of rural people. As Dixon and Hapke (2003) have commented, with particular reference to US farm policy, 'for rural geographers, the examination of agricultural legislation is crucial, because it has such an extensive impact upon the lives not only of farmers but also of rural residents, migrant workers, consumers, businesses at home and abroad, and a host of other groups' (p. 143).

The actions of the state in regulating the rural are not neutral or objective. First, the state will act in what are perceived to be the interests of the larger territorial entity for which it is responsible, such that the rural is always understood to be part of a wider regional or national economy, society or environment. In particular, the state in capitalist societies will typically act in the interests of capital, and will make decisions impacting on rural economies in line with this imperative. The holistic perspective of the state can mean that it adopts positions in rural policy that are perceived by rural residents to reflect urban interests and discourses over rural interests. Second, state policies are informed by political ideologies, which present normative models for the functioning of the economy and for social relations, and which set the parameters for legitimate state action. For example, social democratic ideology holds that the state should intervene to control the excesses of capitalism and promote a more equitable society; whilst liberal ideology limits the scope of the state and contends that the economy functions best when it is driven by market forces.

Prevailing ideologies can switch with a change of government (although more commonly there is a degree of ideological consensus between major political parties in a country), with consequences for the detail of rural policy.

Third, state actors are constantly subject to lobbying by various campaign groups, representing different sectional interests or different views of the rural. Rural policy-making hence proceeds by negotiation with diverse groups, including notably farm unions, producer associations, conservation societies, environmental and animal welfare pressure groups and rural community movements. These groups have varying degrees of access to and influence with the state. A degree of stability is achieved in some contexts by the formation of exclusive 'policy communities' that grant high-level access to a limited number of groups in return for ideological consensus; but in other contexts rural policy-making is more fluid, with competition for influence between competing interest groups in an 'issue network'. As such, the regulation of the rural by the state is a dynamic, contested and sometimes contradictory process.

The next section of this chapter explores in more detail the practice of making and implementing rural policy, with particular emphasis on the significance of the initial stages of defining and describing the rural, which constitute the 'political construction of the rural'. The following two sections then proceed to focus on two of the major concerns for the state in governing the rural: regulating the rural economy and regulating the rural environment. The first discussion highlights the transition of state approaches to the rural economy with the ideological shift to neoliberalism and the consequences for rural communities; whilst the second discussion examines the competing demands on the state in terms of the regulation of the rural environment, especially between conservation and the exploitation of natural resources.

CONSTRUCTING RURAL POLICY

The political construction of the rural

The state's activity in regulating, or governing, the rural necessarily starts with the discursive process of imagining and documenting the rural in order to construct it as an object of governance that can be known and therefore engaged with. In other words, 'the government of [rural] territory entails somehow *knowing* that territory; it requires an understanding of what territory consists of, and what the objectives of government should be' (Murdoch and Ward, 1997: 309). The first step in analysing rural policy, therefore, is to interrogate the discursive assumptions and representations of the rural that underpin policy formulation, which Richardson argues 'means focusing on how rural spaces are constructed within the policy process' (Richardson, 2000: 55). This 'political construction of the rural', consists of four stages of describing the rural, identifying key problems, establishing legitimacy to address these problems, and producing and implementing policies (Woods, 2008a) (see Table 8.1).

These different stages are followed when there is a distinct shift in the orientation of rural policy, as in England in the late 1990s, when the newly elected Labour government sought to reframe English rural policy in response to opposition to its policies on hunting and agriculture. It first set up a unit to collect and analyse statistical data comparing rural and urban areas, and commissioned a 'rural audit' of key issues from leading academics. These reviews presented a very different picture of rural England than that portrayed in conventional agricultural statistics, and that mobilized by political opponents which associated rurality with farming and traditional country pursuits. Instead they emphasized problems of social exclusion, poor housing and access to services – prioritizing political issues that reflected the Labour government's strengths, thus helping it to shed its perception as an 'urban government' and to establish its legitimacy to govern rural England (Woods, 2008a). This reframing of the rural as a political space allowed Labour to propose a more socially focused rural policy (although its implementation was disrupted by an epidemic of Foot and Mouth Disease in 2001, which dragged attention back to agriculture).

More commonly, however, the repetition of standardized statistical and discursive representations of the rural over time reinforces the established political construction of the rural, militating against radical change. In the United States, for instance, the primacy of agriculture in rural policy can be traced to the way in which the unsettled expanse of 'rural' America was described and mapped as a canvas for agricultural cultivation, surveyed and divided into plots for individual settler-farmers, thus creating an agricultural landscape founded on individual enterprise (Opie, 1994). This act of description converged with a Jeffersonian discourse of agrarianism

Table 8.1 Stages in the political construction of the rural, with examples from England, late 1990s (after Woods, 2008a)

Stage	**Process**	**Examples from English rural policy, late 1990s**
Description	Describing the social, economic and environmental characteristics of rural space, through statistics and maps, and through the representation of the rural through photography, prose, text and anecdote.	Reports from the Cabinet Office, Countryside Agency, and the Rural Group of Labour MPs emphasizing the decreasing importance of agriculture, the significance of social and economic issues, and comparing rural and urban areas.
Problem Identification	Identifying from the above description the major problems of rural people and rural areas needing to be addressed by government, setting priorities for rural policy.	Analysis in reports identifying health, education, employment, housing, transport and access to services as key issues for the rural population (not farming and hunting).
Establishing Legitimacy	Demonstrating the competence and mandate of a state institution or of a political party to propose solutions to rural problems and to intervene in the areas identified.	Speeches by Prime Minister Tony Blair and other government ministers, 1999–2000 which stressed support for Labour in rural areas; the shared problems of rural and urban areas; and the government's 'one nation' philosophy.
Delivery	Proposing and implementing a programme of rural policies.	Proposals set out in the 'Rural White Paper', *Our Countryside: The Future* (November 2000), and implemented through legislation and government programmes.

that celebrated the yeoman farmer as the embodiment of the American character, and with the interest of the federal government in ensuring the domestic supply of agricultural produce for food, fuel and manufacturing, to produce a political consensus in which the primary objective of rural policy was accepted as being to support the endeavour of independent farmers (Woods, 2010b).

The United States Department of Agriculture (USDA) was consequently one of the first federal government departments to be established, and rapidly became one of the largest. Periodically, in response to economic and environmental crises, the USDA has engaged in initiatives for broader, non-agricultural rural development, yet as Browne (2001) observed:

> These efforts had to be small because any extensive policy initiatives would have acknowledged farm policy failure and brought resulting policymaker scepticism about this assistance. Thus, the limited size, funding, and means of evolution of rural programs always made clear that this sort of farm policy accommodation did not bring forward a decisive national policy for those left behind by farming.
> (Browne, 2001: 43)

Indeed, agricultural primacy has survived in spite of the circulation of alternative discourses of rural America, some of which have successfully captured parts of the state apparatus – for example, aspects of conservation policy (Woods, 2010b). Opie (1994) rightly suggests that through the dominant agrarian discourse, 'Americans have assigned "duties" and roles, even a moral imperative, to American farmland and American farmers that have persistently made both vulnerable' (p. xiii), yet the favoured response has been to deal with vulnerabilities through internal adjustments to agricultural policy – for instance, by shifting the emphasis over time from individual farmers to agri-business.

Agricultural primacy therefore stems not only from the discursive construction of rural policy, but also from the ability of the agricultural lobby to ensure that agricultural representations of the rural remained foremost in policy-making, even in the context of decreasing farm employment. The influence of the agricultural lobby, and particularly farm unions, has several explanations, including the embeddedness of farm unions in rural areas as multi-functional social and political organizations; the forging of close alliances between farm unions and political parties; and the absence of competitor organizations at the national scale representing alternative non-agricultural rural interests (Sheingate, 2001; Woods, 2005b). The precise configuration of these factors varied reflecting particular national contexts. In Australia, for example, the Country Party became the vehicle for farmers' political representation and succeeded in monopolizing control of the Agriculture Ministry for much of mid-twentieth century; in Japan, farmers' cooperatives were incorporated as part of the political machine of the Liberal Democratic Party; whilst in France, the dominant farmers' union, the *Fédération nationale de syndicates d'exploitants agricoles* (FNSEA), accommodated right- and left-wings that ensured that it had links with both conservative and socialist political parties (Sheingate, 2001).

Most important, however, was the incorporation of farm unions and producer groups into policy-making structures by states anxious to guarantee food supplies in the post-war period. In countries such as Britain and the United States, as well as in the European Community, agricultural policy became the preserve of a close-knit 'policy community' composed of the agricultural ministry and the large farm unions (Smith, 1989). Within this policy community, farm leaders enjoyed high-quality access to ministers and civil servants and were routinely directly involved in policy-making. In return, farm unions assisted the implementation of policy, including the collection of agricultural statistics which reinforced the description of the rural as an agricultural space. The participants in the policy community shared in a productivist consensus (see Chapter 3), and actors who may have presented an alternative representation, including environmental, consumer and animal welfare groups, were excluded. As such, agricultural policy communities only received representations of the rural that concurred with their pre-conceptions and were thus affirmed in the correctness of their approach.

The policy community model delivered a stable political environment for the development of productivist agriculture in the post-Second World War era, but it was exposed by its failure to respond to changing political pressures. By the 1980s, growing public recognition of the problems of productivist agriculture (see Chapter 3) was putting pressure on governments for policy reform. At the same time, the rise of 'neoliberal' political ideology was promoting a new mode of 'governmentality', or a new way of doing rural policy, in which the state-centric, interventionist approach of the policy communities did not fit.

Governmentality and governance

The re-appraisal and re-orientation of rural policy in many countries at the turn of twenty-first century has been part of a wider shift in the mode of governmentality in Western liberal democracies. Governmentality is the process by which the state fixes questions about who should be governed, by whom, to what ends and by what means (Cheshire and Woods, 2009; Murdoch, 1997b). As such, the political construction of the rural described above forms part of the practice of governmentality, providing a means by which the state '"problematizes" life within its borders and seeks to act in response to the resulting "problematizations"' (Murdoch and Ward, 1997: 308). However, governmentality also refers to broader questions about the nature of the state and the parameters of legitimate state action, as well as to the technologies that enable the state to construct knowledge about its realm and to implement policies and exercise power over the population (see Box 8.1).

Box 8.1 GOVERNMENTALITY

Governmentality refers in broad terms to the ways in which society is made governable. The concept originates with the French philosopher Michel Foucault, who used it in the context of interrogating the 'problematics of government' – questions about who should be governed, by whom, to what ends and by what means (Foucault, 1991). In a separate body of work, Foucault also uses governmentality to refer to the development of new ways of governing associated with the extension of state activity into new areas concerned with social behaviour and wellbeing. This, Foucault (1978) suggested, required the invention of new types of power and with them, new technologies of government for the calculation, audit and inscription of the population.

The concept of governmentality was subsequently developed by theorists including Mitchell Dean (1999) and Nikolas Rose (1996) as a framework for understanding the evolution of the liberal state, and in particular, for exploring the question of 'how is it possible to govern a "free" society?' (Murdoch, 1997b: 109). In so doing, they focus on the rationalities through which the state reflects on its own being and purpose, and through which it problematizes its realm as an object of governance. The resulting 'modes of governmentality' are temporary fixes that evolve in response to changing social and economic circumstances and with the waxing and waning of political ideologies.

Rose (1996) accordingly distinguishes between two modes of 'managed liberalism' and 'advanced liberalism'. Managed liberalism, which prevailed in the post-Second World War period, held that the state has a function in managing liberal society in order to restrain the excesses of unchecked capitalism. This was primarily organized at the scale of the nation state, and in social democratic regimes additionally involved the development of welfarism based on a common set of social entitlements for national citizens (access to education and health care, unemployment benefits, state pensions, etc.). As such, Rose refers to this as 'governing through the social'. Advanced liberalism, in contrast, has gained prominence since the 1980s and seeks to limit the legitimate scope of state activity. It instead involves a rationality of 'governing through communities', with responsibilities transferred to citizens and communities who are expected to 'help themselves', with the state 'governing from a distance' as a facilitator and monitor. In its emphasis on limiting state bureaucracy, but its acknowledgement that some degree of regulation is required in order to enable market forces to function, there are strong correspondences between 'advanced liberalism' and the concept of 'neoliberalism' that has been developed independently (see Box 8.2).

Further reading: Murdoch (1997b), Rose (1996).

The practice of governmentality is not fixed, but has evolved over time in response to changing social and economic circumstances and under the influence of different political ideologies. The dominant approach to governing the rural in mid-twentieth-century liberal democracies – with an emphasis on support for agriculture delivered through state intervention and controlled by an exclusive policy community – accordingly reflected the broader dominance of managed liberalism as the prevailing mode of governmentality. Technologies such as the collation of agricultural, environmental and population statistics, and various forms of cartography, were employed to present the rural as a single quantified national unit that could be viewed and understood by a small policy community (Demeritt, 2001; Murdoch and Ward, 1997). The policy community in turn sought to manage the 'national countryside' as a coherent unit, making decisions that were implemented in a top-down fashion through a hierarchical structure centred on the relevant government department. This way of operating helped to protect agricultural policy from external interference, but it was also replicated in other, parallel, 'policy silos' relating for instance to conservation, planning and economic development.

Managed liberalism permitted the rural to be regulated in the 'national interest'. At times it subjugated the rights of individual rural citizens and businesses to the perceived collective interest, but it also accorded to rural citizens the right of equitable treatment from the state regardless of geographical location. As such, managed liberalism saw the rolling-out of public services in rural areas, which additionally helped to support economic development and to encourage counterurbanization. Additionally, the provision of state benefits for the unemployed and deprived removed the burden of paternalism from individual landowners and employers and facilitated the rationalization of the farm workforce and the reconfiguration of rural labour relations.

However, managed liberalism was criticized from the political right for over-stretching the reach of the state and for its cost and bureaucracy. The 1980s saw the election of 'new right' governments in countries including Britain and the United States that adopted a new mode of governmentality, 'neoliberalism' (also referred to in some governmentality literature as 'advanced liberalism'), in which the activities of the state were 'rolled back' through deregulation and privatization (Rose, 1996). The application of advanced liberalism to rural policy has varied in form between countries, and is discussed further in the next section with respect to regulating the rural economy. However, three key characteristics can be identified in relation to the way in which rural policy is constructed and framed.

First, in shifting emphasis from 'governing the social' to 'governing through communities', advanced liberalism is able to recognize the differentiated character of the contemporary countryside and argues that the nation state is not best placed to identify and respond to the resulting policy challenges (Murdoch, 1997b). Second, 'governing through communities' also involves a switch from a vertical perspective, in which policies are delivered top-down from the national government within demarcated 'policy silos', to a horizontal perspective in which an integrated view of the different policy fields impacting on a particular rural community or territory is required. Third, in reducing the activities of the state, advanced liberalism has transferred part of the responsibility for governing the rural to communities, private and voluntary sector stakeholders, and individual citizens, who are encouraged to 'help themselves' (Woods and Goodwin, 2003).

These themes were articulated, for example, in the 'Rural White Paper' published in England in 1997, which marked the first comprehensive statement of government rural policy (as opposed to agricultural policy) in Britain since the Second World War. As Murdoch (1997b) observed, in the document:

> we see how the countryside is, firstly, represented as consisting of small, tightly knit communities. These communities, it is then asserted, are fully able to 'help themselves' and the government proposes ensure that more responsibilities are devolved to the local level. In this way it is hoped a circle can be squared: all the residents of rural areas can have access to basic levels of service and amenity as long as they provide more and more of these themselves.
>
> (Murdoch, 1997b: 117)

Advanced liberalism has consequently produced a new structure of 'rural governance', in which the responsibility of governing is shared between the state and a plethora of community and voluntary groups, private businesses and individual citizens, often working together in 'partnerships' (Goodwin, 1998; Woods and Goodwin, 2003). Again, the precise form of this structure varies: in the Netherlands, for example, the emphasis is on inter-sectoral partnerships (Derkzen, 2010), whilst in Australia it has been on community self-help (Cheshire, 2006). A common thread, however, is the proclaimed capacity to act of rural actors. Both rural communities and individual rural citizens are presented as being resourceful, self-sufficient and independent. Communities and citizens that fail to live up to this stereotype, that require the intervention of the state, are hence implied to be somehow not truly rural (Cheshire, 2006; Woods *et al.*, 2007).

Nevertheless, as Murdoch (1997b) notes, advanced liberalism is not about the withdrawal of the state, but rather involves 'government from a distance'. State actors are often the strongest participants in partnerships, and play a leading role in steering governance (Derkzen *et al.*, 2008; Edwards *et al.*, 2001). The state also continues to set the rules of engagement for governance actors, and supplies funding for projects and initiatives through a quasi-market of competitive bidding that has replaced the previous entitlement to nationwide provision (Cheshire, 2006; Warner, 2006).

REGULATING THE RURAL ECONOMY

From the 'agricultural welfare state' to neoliberalism

The construction of rural space as a source of food and natural resources for energy and building (see Chapter 3), has meant that the state has always had an interest in the rural economy. Forests and mines were historically developed and exploited by political leaders and state actors, often for military purposes. Governmental authorities at different scales were also historically active in regulating trade in food, notably through the imposition of tariffs and taxes, and political debates in the nineteenth century over free trade typically pitched export-orientated manufacturers against protectionist farmers and landowners. However, there was little direct interference by the state in the everyday practice of farming, and no safety-net provided by government to help farmers if crops failed or prices were poor.

It was only in the mid-twentieth century that the state in western capitalist countries began to intervene more systematically in the agricultural economy (Table 8.2). The rationale behind this development was two-fold. First, governments had become concerned about food security. Many countries, particularly in Europe, experienced severe food shortages in the period immediately after the Second World War, and projections of a rapidly urbanizing population made the need to guarantee a reliable supply of food even more pressing. Descriptions of the agricultural sector at the time, however, made it clear that farming would need to be extensively modernized and industrialized in order to meet the demand, and indicated that this could not be achieved through private investment from within the sector, which was predominantly composed of small farms.

Table 8.2 Timeline of key developments in agricultural policy and trade

Agricultural Policy	**Trade Policy**	
1862 United States Department of Agriculture (USDA) created		
1914–18 Wartime agricultural committees introduce state intervention in farming in Britain		
	1930 Smoot–Hawley Tariff Act increases tariffs on agricultural imports to US	↑
	1932 Ottawa Agreement creates Imperial Preference System in British Empire	
1933 New Deal policies in US introduce support mechanisms for agriculture		
1946 United Nations Food and Agriculture Organization (FAO) established	1946–48 Global trade talks fail to support liberalization, reinforcing protectionism	PROTECTIONISM
1947 British Agriculture Act		
	1948 General Agreement on Tariffs and Trade signed	
1949 First 'Farm Bill' in US		
1954 US starts to reduce agricultural price supports (further reforms in 1964 and 1973)		
1957 European Economic Community (EEC) created with a Common Agricultural Policy (CAP)	1957 'Common market' for trade created within EEC	
1963 World Food Programme created to distribute emergency food aid		↓

(Continued Overleaf)

Table 8.2 Continued

Agricultural Policy	Trade Policy
1975 Pinochet regime in Chile starts neoliberal reforms	
1983 Australia adopts neoliberal policies, including agricultural deregulation	
1984 Radical neoliberal reforms in New Zealand, including end of farm subsidies	
1984 First attempt at reform of CAP introduces milk production quotas	
	1986 Cairns Group established
	1991 Mercosur established
	1992 AFTA established
	1992 Single European Market created within European Community (later EU)
	1994 NAFTA established
	1994 GATT talks reach an 'Agreement on Agriculture'
	1995 World Trade Organization (WTO) established, replacing GATT
2003 CAP reforms 'decouple' farm subsidies from production in EU	2003 Disagreement over agricultural trade causes collapse of WTO summit in Cancún
	2005 EU, US and Japan agree to phase out direct subsidies for agricultural exports at WTO meeting in Hong Kong

PROTECTIONISM

NEOLIBERALISM

Governments across the global north consequently invested in agricultural modernization by providing grants and loans to farmers and by funding agricultural research. Governments also shouldered some of the risk of innovation through market support mechanisms and incentives for the production of export crops.

Second, though, governments were also concerned to protect the viability of small-scale family farms, drawing on a political construction of the rural that positioned such farms as the cornerstone of rural society and the guardians of the rural environment. In the United States, supply management of agricultural goods had been introduced as part of the New Deal in 1933 in order to guarantee returns to family farmers struggling in the Depression (Winders, 2004). In Europe, the Treaty of Rome that established the European Economic Community in 1957 expressly identified one of the objectives of the proposed Common Agricultural Policy (CAP) as being 'to ensure a fair standard of living for the agricultural community' (Article 33, section 1(b)), and required that account be taken of 'the particular nature of agricultural activity, which results from the social structure of agriculture and from structural and natural disparities between the various agricultural regions' (Article 33, section 2(a)). As implemented, the CAP performed the duty of maintaining the living standard of the farm population through an intervention system in which the EEC (later EU) would intervene to buy surplus agricultural produce at an agreed price, should market prices fall beneath a set threshold. Farmers were hence guaranteed a minimum price for their product. Similar systems were employed in other states, including Japan, Australia and New Zealand.

Through these mechanisms, mid-twentieth-century agricultural policy in developed nations in effect established what Sheingate (2001) calls the 'agricultural welfare state'. Although presented as an economic policy, the system of agricultural subsidies and price supports operated during this era also functioned as a social policy, supplementing market incomes with a form of state benefit and redistributing wealth by using tax revenue to prop up farms that would not otherwise survive commercially.

State intervention underpinned productivism and was tremendously successful in meeting the objective of guaranteeing food security in the global north. Yet, this was achieved at a significant environmental and social cost, as discussed in Chapter 3, and also at considerable financial expense to the state. By 1984, for example, the cost of the CAP had swollen to account for 69.8 per cent of the European Community budget (Winter, 1996). Agricultural policy consequently was targeted by advocates of neoliberalism (see Box 8.2), who argued both that the intervention system represented a wasteful and unjustified example of state interference in the economy and that regulation was constraining the development of a truly efficient agricultural industry.

Box 8.2 NEOLIBERALISM

Neoliberalism is a political ideology that combines classical liberalism with neoclassical economics. As such it argues that wealth generation and social benefits are best delivered through the free market, and that the state needs to guarantee capitalism the freedom to operate. In contrast to strict classical liberalism it does not advocate an entirely laissez-faire position in which the operation of the economy is unregulated, but rather contends a degree of regulation is necessary to ensure that the free market is able to operate unhindered. This regulation need not be carried out by the state, but neoliberalism suggests that the role of the state is to facilitate capitalism and market competition.

The origins of neoliberalism were conceived by economists meeting in the Colloque Walter Lippmann in 1938, and its principles were subsequently developed by Friedrich Hayek. The concept had little political appeal in the post-war period, when a model of Keynesian economics or 'embedded liberalism' predominated in western states involving the regulation of free trade and private enterprise to provide stability within capitalism (Harvey, 2005). However, economic recession in the 1970s generated critiques of embedded liberalism, and neoliberalism was promoted by 'New Right' politicians and economists as a means of cutting state expenditure and bureaucracy, liberating business and delivering freedom of choice to consumers. Neoliberal policies were first implemented in Chile following the Pinochet coup in 1975, and were subsequently adopted by the Reagan administration in the United States, Thatcher government in Britain, and in most other capitalist states.

The key features of neoliberalism include trade liberalization and the liberalization of foreign direct investment; deregulation; the privatization of state assets; reduced public spending and an emphasis on pro-growth investment; tax reform; liberalization of financial markets; and discipline in fiscal policy. Neoliberalism is particularly associated with economic globalization and the removal of barriers to global free trade.

> The implementation of neoliberalism has been strongly contested in both the global north and the global south. Critics argue that neoliberalism concentrates wealth and extenuates inequality, and that by privileging market forces over all else, it disregards the social, cultural and environmental impacts of capital accumulation and encourages labour exploitation and environmental degradation.
>
> Further reading: Harvey (2005).

The gradual retreat of the state from agricultural regulation started in the United States as early as 1954, when levels of price support were reduced, with further reforms in 1964 and 1973 introducing greater flexibility, removing constraints on production and making the system more 'market-orientated' (Winders, 2004). Significantly, Winders argues that these reforms reflected not the weakness of the US agricultural lobby, as some commentators have suggested, but rather represented 'a policy favored by important segments of agriculture' (p. 468). Deregulation polarized the American agricultural sector, benefiting large, export-oriented corn producers, but disadvantaging the cotton and wheat industries, and small family farmers. Combined with high interest rates and poor yields, the end of price supports contributed to the US 'Farm Crisis' in the 1980s, when over 235,000 small farms failed, consolidating the dominance of agri-business in American farming.

More dramatic neoliberal reforms to farm policy were implemented in Australia and New Zealand during the 1980s. Both countries had developed large agricultural sectors based originally on exports to Britain. As this trade diminished, especially following Britain's accession to the European Community in 1972, efforts were initially made to encourage diversification through targeted subsidies (Cloke, 1989b), yet with the agricultural economy closely tied to the national economy in these countries, critics argued that more radical restructuring was necessary. Advocates of neoliberal reform again included key figures within the agricultural sector, with economists in Australian farm organizations, for instance, 'playing a role in swinging farm leadership away from a handout mentality towards a concentration on reducing costs including lower tariffs, structural reforms to the waterfront, shipping and transport systems and industrial relations' (Connors, 1996: 57).

In New Zealand, neoliberal reforms were introduced following the election of David Lange's Labour government in 1984. As part of a wider package of radical reforms across the entirety of economic policy, Finance Minister Roger Douglas incrementally deregulated provision for agriculture, including the withdrawal of price supports, the phasing out of various targeted subsidies, and the end of special terms for farmers on loans and taxation (Cloke, 1989b). As Smith and Montgomery (2003) observe, 'by the end of 1985, there had been an almost total dismantling of all the protection secured by farmers under a succession of earlier governments' (p. 107). The contribution of state subsidies to agricultural income was slashed from 33 per cent in 1984 to less than 2 per cent in 2003 (*Ibid*), and by 1993, 'New Zealand agriculture could be characterized as having moved from a relatively high income, protected, low risk environment, to a low income, unprotected environment in which the farmers themselves now carried the primary risk' (Smith and Montgomery, 2003: 109).

The immediate impact was a reduction in farm incomes, especially for hill farms, and an increase in farm debt (Cloke, 1989b). The number of farms that were forced out of business is disputed (Smith and Montgomery, 2003), but for those compelled to leave farming the experience was often traumatic (Johnsen, 2003). Yet, more broadly, New Zealand agriculture adjusted to the new system. Farmers diversified into new sectors, notably horticulture and viticulture, and innovated with new processing methods and new export markets, especially in dairying (Smith and Montgomery, 2003). Despite initial opposition by farmers to the reforms, a discursive consensus emerged between the government and the main farm union, the Federated Farmers of New Zealand, that held the reforms to have been a success, and which evangelized the diffusion of neoliberalism to other countries (Liepins and Bradshaw, 1999).

In Australia, neoliberal reforms have been introduced more gradually by both Labour and Liberal governments since 1983, and have included the dismantling of systems of guaranteed prices, intervention buying and production quotas; the establishment of a National Competition Policy to encourage more competitive prices, and the removal of internal barriers to competition within Australia; and the aggressive promotion of international free trade. As in New Zealand and the United States, deregulation has produced a realignment of agriculture in Australia towards larger, export-oriented, super-productivist producers at the expense of family farms, and a restructuring of the agricultural geography (Argent, 2002). Cocklin and Dibden (2002), for example, note that the deregulation of the Australian dairy sector in 2000 triggered the closure of many farms (with a 20 per cent reduction in dairy farm numbers in even the strongest dairying state, Victoria, between 2000 and 2004 (Dibden and Cocklin, 2005)), yet they also argue that neoliberal discourse has achieved a hegemonic status in Australian rural policy that precludes consideration of opposition expressed in terms of farm closures' impact on rural communities.

Against this international trend, the European Union stands out for its sceptical position on neoliberalism in agricultural policy. It has instead embraced the concept of multifunctionality, discussed in Chapter 3, and attempted to shift the CAP away from production subsidies towards support for agri-environmental and rural development initiatives (Potter and Tilzey, 2005). As such, the EU argues that it can deregulate agricultural markets and facilitate competition, whilst also supporting the social and environmental benefits of small-scale agriculture. However, the CAP continues to be controversial and expensive, still comprising nearly half of the EU budget, and the success of the EU's strategy in promoting multifunctionality as an alternative to neoliberalism rests on acceptance by the World Trade Organization, indicating the growing importance of the global scale in the regulation of the rural economy.

Neoliberalism and global trade

Although neoliberal policies have been introduced by national governments, the logic of neoliberalism is to eliminate national boundaries and differences as barriers to the operation of the market, and move towards a single global free market. The globalization of free trade has impacted on all sectors of the rural economy – mining and forestry are now in effect global industries, and manufacturing in rural areas of developed countries in Europe and North America has been squeezed as production has been relocated to cheaper locations, particularly in Asia – but the most contentious issues concern agriculture, not least because of the way in which farming has traditionally been constructed in terms of national identity and national interest. As Peine and McMichael (2005), comment:

> Agriculture is associated, and often represented, as a national resource, but it has been increasingly constructed as a global economic value, via the discourse of market rule and comparative advantage. National agricultures, therefore, are brought into competitive relation to one another via market rule, which, we argue is anything but rational in its substantive social and cultural consequences.
>
> (Peine and McMichael, 2005: 19–20)

Neoliberal globalization replaces national government of agriculture with market governance, the rationality of which dictates that only questions of economic performance and regulation are relevant. Issues about the social, cultural and environmental impacts of economic liberalization are not considered, as Peine and McMichael imply above, and neither are questions about labour relations and exploitation and human rights. In this way, neoliberal market governance can be presented as a decoupling of agriculture from the rural. In contrast to previous modes of national regulation, in which agricultural was discursively positioned as core to rural economies and societies, neoliberal market governance simply treats agriculture as an industry, with no reference to its spatial setting.

Institutionally, the transition to neoliberal market governance has limited the capacity of nation states to act with regard to agriculture in three ways. First, there has been a privatization of certain governance functions, especially with respect to product quality. Quality within the agri-food system is increasingly regulated through various labelling and third-party certification schemes, as well as by the quality controls imposed by major transnational food processors and retailers through direct contracting (Busch and Bain, 2004). Second, states have been obliged by international agreements to regulate their internal agricultural economies in particular ways and to adopt policies that enhance competition. Third, new supranational bodies have been created to police and promote global trade and to penalize perceived anti-competitive behaviour, most notably the World Trade Organization (WTO). Accordingly, it has been argued that neoliberalism has led in practice not to the deregulation of agriculture, but to the re-regulation of agriculture with a market-oriented rationality (Busch and Bain, 2004; Higgins and Lawrence, 2005).

However, the development of global free trade in agriculture has been piecemeal, with unrestricted trade largely confined to numerous regional free trade blocs, with varying degrees of liberalization and integration. These include the single market of the European Union, at the most integrated extreme, as well as diverse 'free trade areas', 'customs unions' and 'common markets', including AFTA (South East Asia Free Trade Area), CAFTA (Central American Free Trade Agreement), MERCOSUR (South American Common Market Agreement), NAFTA (North American Free Trade Agreement) and SADC (Southern African Development Community). Many of these areas have progressively expanded in size, and there has been a degree of up-scaling with, for example, CAFTA and NAFTA being subsumed into a new Free Trade Area of the Americas (FTAA).

At the global scale, negotiations on the liberalization of agricultural trade have preoccupied the WTO for over a decade (Narlikar, 2005). An Agreement on Agriculture was concluded at the end of the 'Uruguay Round' of WTO negotiations in 1994 that set out the principles for liberalization. Under the agreement, 'policies that interfere with trade (including interference with production, price, imports or exports) are subject to reduction and eventual elimination under WTO rules' (Peine and McMichael, 2005: 24). However, discussions on further measures have faltered several times at Geneva (1997), Doha (2001), Cancún (2003) and Hong Kong (2005). Radical liberalization has been pushed by the Cairns Groups of agricultural exporting nations, including Australia, Argentina, Brazil, Canada, Chile, New Zealand, South Africa and ten others. The United States has adopted a more equivocal position, and the European Union has sought acceptance for environmental support under its model of multi-functionality (opposed by the Cairns Group) (Potter and Burney, 2002). At the same time, WTO meetings have routinely attracted large protests, including farmer activists from both the global north and the global south. Peet (2003), for example, describes the rally outside the Geneva meeting in 1997 as comprising 'farmers who believed that with the removal of trade barriers multinational corporations would take over their markets and lands, workers protesting about job losses, and consumers concerned about harmful products such as tobacco and genetically engineered foods' (p 192). As such, the WTO has become a key discursive site in which the future regulation of the rural economy is contested.

Advocates of global free trade argue that it will benefit developing countries most. McCalla and Nash (2007), for example, suggest that 'recent estimates are that developing country income would be some 0.8 per cent higher in 2015 than it would otherwise be if all merchandise trade barriers and agricultural subsidies were removed between 2005 and 2010, with about two-thirds of the total gain coming from agricultural trade and subsidy reform' (p. 3). Yet, this analysis is an example of the decoupling of agriculture from the rural; it considers only the net economic effect, and pays no regard to the social, cultural or environmental impacts. Grinspun (2003), in research on Uruguay, puts the alternative case that:

> the expansion of global markets for rural products, along with the aggressive introduction of industrial methods and export orientation to rural economies, are undermining smallholder and subsistence farming as well as non-agricultural family enterprises. They are also creating new barriers to the efforts of small-scale rural enterprises to diversify their agricultural and non-agricultural activities.
>
> (Grinspun, 2003: 48)

Studies of the impact of regional trade agreements have indicated that the removal of trade barriers detrimentally affects rural economies and communities in at least three ways. First, farmers are forced to compete in domestic markets with imported agricultural produce, which may have lower production costs. Competition hence typically drives down prices, and slashes the income to farmers. Following the inception of NAFTA in 1994, corn prices in Mexico fell by 70 per cent, with 1.75 million peasant farmers forced to leave the land (Peine and McMichael, 2005). Second, transnational corporations assume a greater presence in agricultural markets, reconfiguring local supply chains. Nestlé and Parmalat, for example, selected Uruguay as a base for expanding activity in the MERCOSUR free trade area, but their capture of the dairy industry disrupted markets and squeezed incomes for dairy farmers in western Uruguay (Grinspun, 2003). Third, rural economic and social relations are also restructured by internal reforms that governments are obliged to introduce as part of trade agreements, such as the phasing out of subsidies, land reform or incentives for agri-business. McDonald (2001), for instance, describes the implementation of post-NAFTA reforms in the Mexican dairy industry, which he argues re-shaped rural power relations and increased inequalities within rural society.

Neoliberalism and depeasantization in Chile

The long-term impact of neoliberalism on rural areas in the global south is illustrated by the case of Chile, which was the first country to adopt neoliberal policies. As Harvey (2005) records, a situation existed in the early 1970s where:

> the policies of import substitution (fostering national industries by subsidies or tariff protections) that had dominated Latin American attempts at economic development had fallen into disrepute, particularly in Chile, where they had never worked that well. With the whole world in economic recession, a new approach was called for.
>
> (Harvey, 2005: 8)

A military coup in 1975 and the installation of the Pinochet dictatorship provided the opportunity for the implementation of experimental, neoliberal economic policies intended to 'reverse the collectivization of previous periods, harness market forces in rural land and labour, and return power and foster capitalist potential in the landlord class' (Murray, 2006a: 654). Policy instruments included returning expropriated land, wage cuts, withdrawing subsidies to small farmers and ending price controls on food. Foreign investment was encouraged, and export industries promoted. In particular, an emphasis was placed on the development of 'non-traditional agricultural export' (NTAX) crops, especially fruit. The Chilean fruit-growing sector had been negligible in 1970, but expanded to become the country's second main export (after copper) by the 1980s, and by 2003 was worth US$1.7 billion per year (Murray, 2006a).

A feature of the initial reforms was the creation of a *parcelero* sector of small farmers holding between 5 and 20 hectares, aimed at stimulating rural capitalism. This stage of *re-peasantization* was followed by a process of *peasant incorporation*, in which small farmers were drawn into the expanding export industry, which in turn was superseded by *proletarianization*, as the *parcelero* sector was subsumed by corporate interests from 1990 onwards. In the locality of El Palqui, in the Guatulame Valley, for example, only 15 per cent of the original *parcelero* farmers who benefited from the land reforms were still operating in 2005 (Murray, 2006a). Meanwhile, five export companies had between them acquired 45 per cent of land in the area. As such, 'El Palqui is now an agrarian context where a small number of large capitalist farmers and agribusiness enterprises control the majority of the land, water and labour-power in the settlement' (Murray, 2006a: 668).

Peasants and small farmers have hence been separated from the means of production in a process of depeasantization. Farmers who have lost land have been proletarianized, as they have become labourers for corporate agriculture, or migrated to urban areas. In addition to those farmers who lost their land, others have been 'semi-proletarianized', retaining access to land, but becoming 'entirely dependent on agribusiness to the extent that they effectively became tied, yet generally unwaged, labour employed by that company' (Murray, 2006a: 650). Consequently, whilst 30 years of neoliberalism in Chile (in moderated form since the overthrow of the Pinochet regime), has brought significant growth to the economy, this 'success' has 'been built on and has perpetuated a deepening socio-economic differentiation in the countryside, with rising relative poverty, increased peasant marginalization, and a worsening income distribution across the country as a whole' (*Ibid*: 653).

REGULATING THE ENVIRONMENT

Conserving the countryside

Alongside policies to support the functioning of a production-oriented rural economy, and particularly an agricultural-centred economy, the state has also responded over the course of the last century and more to a second imperative, the demand for state intervention to regulate and protect the rural environment. This alternative political construction of the rural stems from romantic and pastoral representations of the rural that were popularized in the nineteenth century, and from the realization that such representations captured a rural landscape that was increasingly threatened by cultivation, industrialization and urbanization (Bunce, 1994) (see Chapter 2). Part of the response was the organization of private initiatives to safeguard valued rural landscapes, such as the founding of the National Trust in Britain in 1895 as a charitable body that bought areas of rural land to protect and manage them for the nation. However, it was also argued that the state itself should have a role in regulating the development of rural land. This suggestion represented a significant proposed extension to the reach of the state at the time, beyond the economic and social spheres to the environment. It also potentially conflicted with the interests of capital accumulation, which was the primary concern for the state. As such, the state's intervention in regulating the rural environment took the form of spatial regulation, imposing spatially differentiated controls on the development of land and designating specific areas for special protection, but generally leaving the operation of agriculture and other primary industries alone in the majority of the countryside.

In North America, the initial pressures for state intervention focused on the protection of wilderness areas, which were gradually being eroded by the advance of European settlement, cultivation and exploitation by mining. One of the first people to call specifically for state intervention was the artist George Catlin, who proposed the setting aside of landscapes of natural grandeur 'by some great protecting policy of the government' to create a 'nation's park containing man and beast, and in all the wild and freshness of their nature's beauty' (Catlin, 1930: 294–95). The reference to a 'nation's park' was deliberate and became a key element in the discursive strategy of the preservationist lobby, positioning the protection of landscapes of distinctive American natural and cultural heritage as being in the national interest. This discourse was reinforced by Frederick Jackson Turner's 'frontier thesis', which argued that the preservation of a wild frontier was essential for American democracy, helping to define the national character, and acting as a safety valve for urban problems (Turner, 1920).

The first significant example of state intervention to protect a rural landscape came in 1864 when Mariposa Grove (later Yosemite National Park) was presented to the Californian state government to be managed for public use and recreation. As such, whilst Yosemite was identified as a 'national treasure', its management by the state owed more to the model being developed by municipal governments at the time of the public provision of pleasure parks for amenity purposes (see Jones and Wills, 2005). However, the precedent of combining conservation and recreation was set and continues to be a feature of most national parks. Further conventions were established with the formation of the first true 'national park' at Yellowstone in 1872. Many of these conventions were accidental – Yellowstone was administered by the federal government because it was outside any existing state at the time; it was publicly owned because no claims of private landownership had been made; and it was unpopulated because it had not been settled – but came to be recognized as the Yellowstone model.

The transposition of the Yellowstone model to more populated areas has consequently presented difficulties. The creation of Great Smoky Mountains National Park in North Carolina and Tennessee in 1934, for example, required the resettlement of settler farmers and loggers (the indigenous Cherokee had already been evicted), and the more recent establishment of national parks in many areas of the global south has frequently been accompanied by the displacement of resident populations – contrary to Catlin's vision which was to protect both wildlife *and* the indigenous human culture. Perhaps best known is the resettlement of Maasai people from the Serengeti National Park in Kenya in 1959, which was followed by the further restriction of hunting and cultivation by the Maasai as conservation areas were subsequently extended in the surrounding region (Monbiot, 1994). Brockington and Igoe (2006) identify over 240 cases of forced eviction of indigenous peoples from protected areas in Africa, Asia and Latin America, including in the creation of the Limpopo National Park in Mozambique in 2001, where over 31,000 people in 53 villages were moved. Even where resettlement has not been enforced, the establishment of protected areas in the global south has frequently imposed restrictions on the traditional use of rural land and resources by indigenous peoples (Box 8.3). In contrast, many European countries have adopted a modified model of national parks in which parks continue to be populated and privately owned, necessitating a greater degree of negotiation between the state and the local population.

Protecting rural space

In late nineteenth- and early twentieth-century Britain, the greater concern of preservationists was with protecting rural landscapes from urban encroachment. In the preservationist discourse, the essence of rural England stemmed its opposition to the urban. As such, the greatest threat to the countryside was perceived to come from the blurring of the urban–rural divide, as articulated by the planner Thomas Sharp:

> From dreary towns, the broad, mechanical, noisy main roads run out between ribbons of tawdry houses, disorderly refreshment shacks and vile, untidy garages. The old trees and hedgerows that bordered them a few years ago have given place to concrete posts and avenues of telegraph poles, to hoardings and enamel advertizement signs.

Box 8.3 INDIGENOUS CULTURE AND PROTECTED AREAS IN MADAGASCAR

The Ankarana Special Reserve in northern Madagascar was established in the 1950s, at the end of the French colonial era, to protect the unique rainforest ecosystem from exploitation and deforestation. Management of the area for conservation was subsequently strengthened as part of Madagascar's National Environment Action Plan, developed with encouragement from international donors, including the World Bank. Although the creation and management of the reserve did not involve the forced resettlement of communities, it followed a 'colonial protected area paradigm', 'focusing on the enforcement of boundaries that local people were not allowed to cross' (Gezon, 2006: 37). Later modifications of the approach have attempted to bring local indigenous leaders into the management process, but this has contributed to tensions within local tribes as conflicts between conservation priorities and the needs of local people arise.

The enforcement of the Ankarana Special Reserve has restricted the use of the forest by the indigenous Antankarana people, both for traditional ceremonial and for spiritual purposes, and more prosaically for food and other resources. Gezon (2006) documents one case of a village on the southern edge of the reserve where villagers petitioned the indigenous leader, the Ampanjaka, for permission to source wood for construction from the forest, against an official prohibition. The Ampanjaka, who had been courted by the reserve managers, initially rejected the plea and threatened 'spiritual sanction' against anyone cutting down trees. This injunction was not accepted by village elders, who protested that wood was desperately needed to repair homes. With ties between the Ampanjaka and conservation managers weakened by a change in park leadership, he chose not to assert his authority, allowing an implicit compromise to be reached in which villagers illicitly harvested trees unchallenged. Gezon cites this outcome as an example of the persistence of grassroots movements being able to overcome international prohibitions.

Further reading: Gezon (2006).

> Over great areas there is no longer any country bordering the main roads: there is only a negative, semi-suburbia.
>
> (Sharp, 1932: 4, quoted in Murdoch and Lowe, 2003: 321)

Preservationists called for state regulation in order to check urban expansion, and to separate rural and urban space. In so doing they recognized that liberalism would need to be moderated, Clough Williams-Ellis, for example, stating that 'the choice lies between the end of laissez faire and the end of rural England' (quoted in Matless, 1998: 28). The Council for the Preservation of Rural England (CPRE – now the Campaign to Protect Rural England), was founded in 1926 to advocate a technocratic solution through the utilization of land use planning and development control to maintain the distinctiveness of the rural and the urban. These principles were encoded into the planning system introduced by the 1947 Town and Country Planning Act, 'established around an administrative separation between the urban and the rural, one that functioned to prevent urban sprawl and to safeguard agricultural land' (Murdoch and Lowe, 2003: 322). Key instruments of the system included the de facto nationalization of land development rights in Britain, such that 'planning permission' must be sought for new building or alterations to existing buildings, combined with policies that discriminated against new build in rural areas and essentially precluded new development in 'greenbelts' around major cities (Gallent *et al.*, 2008).

The British planning system has had a major influence in shaping the material form of the contemporary countryside. It has been successful in restricting urban sprawl, protecting pastoral landscapes and maintaining the broad settlement pattern of rural areas. It also contributed to the strong trend of counterurbanization in Britain by limiting the scope for extended suburbanization and forcing urban out-migrants to leap-frog the greenbelt into rural villages. In particular, migrants were attracted to villages in which the idyllized rural character had been preserved by planning controls, and subsequently sought to use the planning system to protect their investment by blocking new development and restricting further in-migration (Murdoch and Abram, 2002). Murdoch and Lowe (2003) describe this as the 'preservationist paradox': 'preservationist groups such as the CPRE find themselves being supported by the transgressors in order to halt further transgressions' (p. 328).

Moreover, the British planning system has not offered any protection to the rural environment from the impact of agricultural practice. Preservationists considered that 'the most certain way of protecting the well-being of rural communities and enhancing the amenity value of the countryside was to retain the existing area of farmland in productive use' (Murdoch and Lowe, 2003: 322). As such, 'the objectives of rural preservationism and agricultural planning could be forged together' (*Ibid*). The planning system consequently included only limited controls on the erection of farm buildings, or changes to the farm landscape. Yet, within decades productivist agriculture had left a mark on the landscape through the removal of hedgerows, in-filling of ponds and ploughing of grassland. There was also increased recognition of the environmental damage caused by industrialized farming through nitrate pollution of watercourses, soil erosion, habitat destruction and the poisoning of wildlife food chains (Green, 1996).

Evidence of these detrimental effects intensified pressure for state regulation of agricultural practice. This was partly achieved by legislation, for example, banning the use of DDT in pesticides, but has mostly been advanced through measures to encourage more environmentally sensitive practice by farmers. These have included various agri-environmental schemes in the European Union, Landcare in Australia, and the Conservation Reserve Program in the United States. Such voluntary initiatives work as a form of 'soft paternalism', trying to induce changes in the behaviour of farmers through incentives rather than compulsion as part of 'governing from a distance'.

The neoliberalization of nature

In the same way that neoliberal thinking has come to dominate state approaches to the regulation of the rural economy, state regulation of the rural environment has increasingly been challenged, re-assessed and re-oriented by the application of neoliberal rationalities. Such adjustments form part of a broader 'neoliberalization of nature', which has reversed the historic project incorporating nature within the concerns of the state (Whitehead *et al.*, 2008). As Castree (2008a) observes, 'the last thirty years have seen an ever greater variety of biophysical phenomena in more and more parts of the world being subject to neoliberal thought and practice' (p. 136). This has been enacted through various mechanisms including the privatization of natural resources, the marketization of environmental phenomena, the deregulation of environmental controls and the re-regulation of nature in ways that facilitate free market enterprise, the adoption of market proxies in the residual public sector, and the construction of flanking mechanisms in civil society, such as private and voluntary sector initiatives that monitor environmental standards in place of the state (Castree, 2008a). Indeed, Castree argues that 'neoliberalism is *necessarily* an environmental project' (2008a: 143), contending that the neoliberalization of nature is concerned with 'fixing the environment' to facilitate capital accumulation in a free market.

The neoliberalization of state regulation of the rural environment has particularly been manifested in three key ways. First, environmental regulations and environmental subsidies have been challenged as 'barriers to trade'. The European Union's support for agri-environmental schemes as part of its agenda of mulitfunctionality, for example, is contested by the Cairns Group in WTO negotiations. Similarly, attempts by the United States to insist that sea turtle excluder devices be fitted to shrimp fishing vessels in countries supplying the US market were challenged by South East Asian nations as a breach of WTO rules (McCarthy, 2004). However, McCarthy (2004) argues that trade agreements have also advanced the re-regulation of nature, by reinforcing new private property relations that have enclosed elements of the rural environment previously considered to be communal property.

Second, state-owned natural resources, many of which were located in rural areas, have been privatized. These include state-owned forestry and agricultural land as well as utilities such as water supply, with implications not only for the ownership and management of rural land, but also for the provision of key resources to the rural economy. Perreault (2005), for example, demonstrates how the granting of private concessions over water supplies in Bolivia (including for export to mining firms in Chile), diverted water from rural communities and compromised the traditional resource rights of peasant irrigation. Where land and natural resources have remained in state control, for example in national parks, 'these resources and ecosystems are to be managed in market-mimicking ways' (Castree, 2008a: 147), leading to the increased commodification of conservation and utility sites for tourism and recreation.

Third, it has advocated that incentives for conservation and environmental good practice in agriculture and land-management should be provided through the market rather than by the state. This approach places a monetary market value on 'ecosystem services' delivered by the rural environment, such are carbon sequestration, pollution removal, habitat provision and flood protection. Robertson (2004), for example, describes the functioning of a wetland mitigation banking scheme in the United States that aims to 'develop a market in privately-owned "wetland ecosystem services", such as duck habitat, flood protection and biodiversity, as a way of achieving the goals of the U.S. Clean Water Act of 1977' (p. 361). The system works through the provision of bank credits to landowners who agree to restore wetland sites, as Robertson (2004) explains for a farm in Illinois:

> by entering into a complex agreement with federal and county regulatory agencies, the banking firm will sell 'wetland credits' to individuals compelled to buy them by those same agencies. Within five years, the production and sale of ecosystem services in the farm field outside of Aurora will have grossed nearly three million dollars.
>
> (Robertson, 2004: 361)

Furthermore, because banks can compete to sell the credits, and can set any price for the credits, it was hoped that the scheme would 'provide the price signals necessary for a real market in wetland services' (p. 363). However, Robertson also identifies problems in the marketization of ecosystem services that complicate the idealized neoliberal model for the process, including appropriate governance and management systems, noting that 'the use of ecosystem science to define ecosystem services in easily measured, abstract units that can be transacted across space (as all commodities must) without losing their value has proven to be very difficult in practice' (p. 362).

As with earlier modes of regulating the rural, the neoliberalization of nature has therefore had both environmental and social impacts in rural areas. Castree (2008b), in a summary of previous studies, shows that these can vary significantly between different cases. In some instances, environmental improvements have been delivered, such as new wetland sites or cleaner water, but there have also been environmental costs including pollution, loss of habitats and intensified land use. Similarly, whilst some neoliberal projects have been associated with democratization and community empowerment initiatives, the neoliberalization of nature has more commonly contributed to increased social inequalities in rural areas, the loss of communal resources, and in some cases protests and civil unrest.

CONCLUSION

The state is a central player in the production and re-production of the rural. It contributes to the discursive construction of the rural, not least through the development and articulation of various political constructions of the rural in which the rural society, economy and environment are described and problematized as objects of governance. The application of policies based on these political constructions has a significant impact in shaping the material geographies of the countryside – for example, by informing the structure of the rural economy, the flow of migration and the protection of the rural environment.

During the twentieth century, the state's engagement in regulating the rural increased in keeping with a dominant mode of governmentality of managed liberalism. In particular, the state responded to agriculturally centred descriptions of the rural economy and society by intervening to subsidize agricultural modernization and guarantee minimum prices to farmers for their produce. However, the rise of neoliberalism towards the end of century started to dismantle this settlement, and has promoted the forging of a global market for agricultural goods. Similarly, the spatial regulation of the rural environment by the state to preserve pastoral or wilderness landscapes, as well as the tentative involvement of the state in regulating the environmental impacts of farm practice, have been challenged and modified by moves towards the neoliberalization of nature, constructing new relations of environmental regulation.

The regulation of the rural is hence a dynamic and contested arena in which the state is increasingly positioned as just one actor alongside various lobby groups and NGOs, supra-national institutions and private regulatory schemes. Moreover, as the state is not monolithic, but comprises different, fragmented agencies and policy circles, different political constructions of the rural may find purchase in different parts of the state at the same time. The interests of economic regulation and environmental regulation, for example, may be contradictory, leading to conflicts as discussed in the next chapter.

FURTHER READING

For more on the political construction of the rural and its articulation in British rural policy, see the introductory chapter of Michael Woods (ed.), *New Labour's Countryside: British Rural Policy since 1997* (2008a). The application of the concept of governmentality to analysis of rural policy and the rural state was particularly developed by Jon Murdoch, with good examples being his article on the 1995 Rural White Paper in England, published in *Area* (1997b), and his paper with Neil Ward on the use of agricultural statistics to construct Britain's 'national farm', published in *Political Geography* (1997). For an introduction to and historical overview of neoliberalism see David Harvey's *A Brief History of Neoliberalism* (2005). Smith and Montgomery's paper in *Geojournal* (2003) is a good review of 20 years of neoliberal reforms in New Zealand agriculture; Warwick Murray provides a critical analysis of 30 years of neoliberalism in Chile in the *Journal of Peasant Studies* (2006a); whilst Busch and Bain's article in *Rural Sociology* (2004) discusses neoliberalism in global agricultural trade in historical context. For an accessible introduction to the workings of the World Trade Organization see Narlikar's *The World Trade Organization: A Very Short Introduction* (2005). The origins of the preservationist movement are described by Michael Bunce in *The Countryside Ideal* (1994), whilst Jon Murdoch and Philip Lowe's paper in *Transactions of the Institute of British Geographers* (2003) is a good source on the CPRE, the establishment of the British planning system, and the 'preservationist paradox'. Two companion papers by Noel Castree in *Environment and Planning A* (2008a, 2008b) provide an excellent, in-depth, review and critique of geographical research on the neoliberalization of nature.

9

RE-MAKING THE RURAL

INTRODUCTION

The purpose of this book has been to explore the production and reproduction of the rural. In so doing it started with the assertion that the rural is an imagined space; the dichotomy of the city and the country, the urban and the rural, may be one of the oldest and most resilient geographical dualisms, but it is nonetheless an artificial construction, as geographers, planners, sociologists and others have found when they have attempted to delineate rural space or to define the essence of rural society (Chapters 1 and 2). The discursive construction of the rural has involved not only the imagined division of space, but also the filling of rural space with characteristics and meaning. These in turn have been enacted through performances that articulate rurality in everyday practice (Chapter 7), regulated and developed through the legislation, policies and activities of the state (Chapters 6 and 8), and converted into a material countryside that is embodied in the rural landscape, the biodiversity of rural areas, the structure of the rural economy, the pattern and form of rural settlements, and the composition and living standards of the rural population.

However, as this book has also shown, there are many different ways in which the rural can be imagined, described, performed and materialized, and as such, arguably many different rurals. The rural is conceived of variously as a resource to be exploited for economic gain (Chapter 3), a site of consumption through tourism and recreation (Chapter 4), a place to live (Chapter 6), and as a vulnerable environment in need of protection (Chapter 8). Neither is there a single author of the rural. The state, the media, corporations, farmers, rural residents, academic researchers, tourists and day-visitors, pressure groups and NGOs, development agencies, investors and speculators and a host of non-human actors – among others – are actively engaged in the production and reproduction of the rural on an everyday basis.

Accordingly, we can re-affirm Murdoch's (2003) observation, noted in Chapter 2, that 'there is no single vantage point from which the whole panoply of rural or countryside relations can be seen' (p. 274). The rural is hybrid, co-constituted, multi-faceted, relational, elusive. Rural geography, as a field of academic study, has taken some time to recognize this condition, moving away from a positivist, empiricist stance that dominated up to the 1980s, searching for the true, real countryside. The introduction of a political-economy perspective in the 1970s and 1980s revealed the political contingencies of rural relations and positioned rural space and the rural economy and society within the wider context of capitalist society; but it has only been with the more recent influence of the 'cultural turn' and of post-structuralist theories that the socially constructed, hybrid nature of the rural has been fully acknowledged and engaged.

This final chapter looks forward to rural futures. It does so by focusing on three arenas in which the rural is being re-made as a multi-authored, multi-faceted and co-constituted space, and which are starting to attract the attention of geographical research. First, it examines the transformation of the rural under contemporary globalization, and in particular the reconfiguration of rural places through the hybrid engagements of local and global actors. Second, it considers the role of non-human agency in disrupting and re-shaping the rural, and the significance of this for rural adjustment to climate change. Finally, it documents the re-assertion of rural identity by new rural social movements in both the global north and the global south, and their contribution to the contestation of the meaning and regulation of rural space in a 'politics of the rural'.

THE GLOBAL COUNTRYSIDE?

Globalization and the rural

A recurrent theme throughout this book has been the significance of globalization as a driver of rural change, or perhaps, more correctly, the significance of diverse globalization processes as drivers of rural change. The variety of contexts in which globalization has been encountered – trade and economic production, tourism, migration, media representations and environmental regulations – points to the multiple character of globalization, corresponding with Steger's (2003) definition of globalization as 'a multidimensional set of social processes that create, multiply, stretch, and intensify worldwide social [and economic] interdependencies and exchanges while at the same time fostering in people a growing awareness of deepening connections between the local and the distant' (p. 13) (see Box 9.1).

Box 9.1 GLOBALIZATION

Globalization is a widespread but loosely used term that can refer to a condition, or a process, or a discourse. Steger's definition, quoted in the main text, identifies globalization as a set of processes, but is useful in the way that it draws out some of the key characteristics of these processes. First, Steger (2003) proposes that globalization 'involves the *creation* of new, and the *multiplication* of existing, social networks and activities that increasingly overcome traditional political, economic, cultural, and geographical boundaries' (p. 9). Second, globalization involves the *expansion* and *stretching* of social and economic relations, activities and interdependencies over increasing distances. Third, globalization involves the *intensification* and *acceleration* of social exchanges and activities, with connections able to be made across increasing distances in increasingly less time and with increasing frequency. Fourth, 'the creation, expansion, and intensification of social interconnections and interdependencies do not occur merely on an objective, material level' (p. 12), but also involve the development of a *global consciousness*, in which people have a greater awareness of the world as a whole, and their place in it.

Globalization can, however, also be understood as a condition of inter-connection and inter-dependency between localities around the world, also referred to as 'globality' (Steger, 2003). The condition of globalization is manifested in rural localities by the presence of global actors such as transnational corporations or organizations, or immigrants or imported technologies, as well as through networks that connect rural localities to distant localities through trade, travel and consumption. Finally, globalization can also be conceptualized as a discourse that presents the existence of a global economy and society as a given and interprets other problems and actions through this lens. Larner (1998), for example, argues that a globalization discourse has been central to the rationality of neoliberalism in New Zealand (see Chapter 8).

There are also different theoretical perspectives on the merits and inevitability of globalization. *Hyper-globalists* (or globalists) see globalization as the natural and unstoppable march of economic integration that has already created a global economy, ushering in a new historical era in which national borders are being dissolved and economic agents organize for competition in the global marketplace. *Traditionalists*, or sceptics, in contrast, argue that globalization is not as advanced as hyper-globalists claim and suggest that globalization has been over-hyped in order to support an imperialistic project of capitalist expansion. *Transformationalists* meanwhile steer a middle route, recognizing that new processes of intense integration and interdependence are occurring, and that these are transforming social, economic, cultural and political relations, but arguing that globalization is incomplete and that its outcomes are not pre-determined (see Murray, 2006b).

Further reading: Murray (2006b), Steger (2003).

For simplicity, the plethora of globalization processes impacting on the rural can be distilled into three broad trends. First, *economic globalization* is having a transformational effect on the economies of rural areas through a number of different but interconnected elements. These include the liberalization of international trade, including in agricultural goods, and the promotion of a global marketplace (Chapter 8); the development of global commodity chains, or global value chains, in which a commodity may be produced in one country, traded in a second, processed in a third, and sold in a fourth (e.g. Barrett *et al.*, 1999 on horticulture; Neilson and Pritchard, 2009 on tea); corporate concentration and the consolidation of transnational corporations and alliances, including in the agri-food sector, forestry and mining (Hendrickson and Heffernan, 2002); foreign inward direct investment, and the consequential vulnerability of rural branch plants to distant corporate decision-making (Epp and Whitson, 2001; Inglis, 2008); and the emergence of new global property regimes, including bioprospecting (the commodification of biological resources by transnational corporations), corporate land investments and the acquisition of land or farming rights in foreign countries by wealthy states with limited agricultural resources seeking to ensure food security.

Second, a *globalization of mobility* has been facilitated by advances in transport and communications technologies and the liberalization of travel and immigration controls. This has contributed to a stretching of phenomena such as tourism, counterurbanization and labour migration to the global scale, with significant implications for many rural localities. The expansion of global tourism has accelerated the commodification of rural landscapes and rural experiences, especially through adventure tourism and eco-tourism (Chapter 4); transnational amenity migration has sought out rural resort locations in both the developed and the global south, and has contributed to the fluidity of rural communities (Chapter 6); and flows of labour migration have both intensified and expanded over wider areas, and have created 'transnational villages' (Chapter 6).

Third, *cultural globalization* has been manifested not only through the convergence and global circulation of media representations of the rural (Chapter 2), but also through a related *globalization of values*, in which there is an increased expectation that the same ethical standards will be applied universally across the world. In a rural context, this has been pursued in particular with respect to animal welfare, including farm husbandry practices and hunting (Chapters 6 and 7), and with regard to conservation and models for environmental protection such as national parks (Chapter 8). Transnational NGOs, such as Greenpeace, Friends of the Earth, PETA and WWF, as well as various international agreements, have played a key role in promoting and policing new global standards, but the assertion of global values has often challenged and constrained embedded local understandings of nature and ways of engaging with nature (Alphandéry and Fortier, 2001).

These processes of globalization are helping to produce a new countryside at the start of the twenty-first century, however, the long-view perspective on the production and reproduction of the rural adopted in this book shows that the influence of global or foreign actors on rural localities is nothing new. Chapter 2, for example, discussed how European ideas of rurality were transported around the world from the fifteenth century onwards, dramatically transforming the rural spaces of the Americas, Africa, Asia, Australia and New Zealand. Similarly, Chapter 3 noted the global character and connections of resource capitalism as it developed in the nineteenth century. What sets the contemporary condition of globalization apart is the totalizing and instantaneous nature of the relations that it embodies. Whereas in earlier eras, transnational networks were formed bilaterally – for example, between colonial powers and their colonies – or targeted specific sites such as mining districts, thus leaving vast swathes of rural space relatively unconnected into global relations, there are now few rural locations that are not integrated in some way into networks and flows that are more or less global in reach. The speed of connection across global distances has also been accelerated. Communication can be achieved immediately with all parts of the world; agricultural goods can be air-freighted to be sold as fresh produce on different continents; tourists can be in remote rural locations on the other side of the world within hours; migrants are able to continue to participate in the community life of their home districts; and the impact of an economic or political crisis in one part of the world reverberates through distant economies. Additionally, contemporary globalization is characterized by the organizing principle of neoliberalism, driving liberalization and moving towards the ambition of a single global market.

This condition has given rise to the construction of global representations of the rural that obscure national differences, the neoliberal vision of global agriculture as a single system of food production being one example. The global circulation of media representations of the rural, combined with the fact that a predominantly urban population now comes to know the rural primarily through the media rather than through direct experience, has similarly promoted a hybridized public image of the countryside in which diverse elements of rurality from different settings are mixed together.

However, the discursive homogenization of rural representations has not been matched by material homogenization. Globalization processes are not creating a single, undifferentiated global rural space, but rather are contributing to re-making rural places in different ways, re-aligning rather than eradicating existing geographies.

Globalizing rural localities

The majority of rural geographical and rural sociological research on globalization to date has focused on broad trends, or on individual industries or commodity chains. In comparison with urban research on the 'global city', there have been few studies of the integrated impact of globalization processes in specific rural localities. As Hogan (2004) has observed, 'there is a discernable privileging of urban over rural in scholarly accounts of globalization' (p. 22). This relative neglect means that our mosaic of understanding of globalization in a rural context continues to be only partially developed:

> the mosaic remains very much a work in progress. Some parts of the picture are considerably clearer and more complete than others; some studies sit as isolated tiles, apart from the emerging tessellation; and the connections between some parts of the image and other parts are as yet unknown. In particular, the mosaic is missing the input of a substantial body of place-based studies – research that might not only adopt an integrated perspective in examining the impact of different forms and aspects of globalization in a rural locality, but that might also explore precisely how rural places are remade under globalization, and start to account for the differential geographies of globalization across rural space.
>
> (Woods, 2007: 490)

The model of the 'global countryside' is presented in Woods (2007) as a framework for developing a locality-based analysis of globalization in rural areas. It is not intended to refer to an actually existing rural territory, at least not at present, but instead is conceived as a hypothetical space characterized by attributes which represent the projected end-point of current globalization processes impacting on rural space (see Table 9.1). As such, the 'global countryside' is intended to highlight the multiple ways in which rural places are restructured through globalization, as well as to expose some of the power relations that are involved in such restructuring. Moreover, as the 'global countryside' does not exist anywhere in its pure and complete form, but is rather always in a state of becoming, the concept emphasizes questions about how globalization works in specific places.

The answers to these questions, Woods (2007) argues, lie in the hybrid constitution of the rural, as constructed not only by both human and non-human entities, but also by both local and non-local actors and forces.

Table 9.1 Characteristics of the emergent global countryside (after Woods, 2007)

1. Primary and secondary sector economic activity in the global countryside feeds, and is dependent on, elongated yet contingent commodity networks, with consumption distanced from production.
2. The global countryside is the site of increasing corporate concentration and integration, with corporate networks organized on a transnational scale.
3. The global countryside is both the supplier and the employer of migrant labour.
4. The globalization of mobility is also marked by the flow of tourists and amenity migrants through the global countryside, attracted to sites of global rural amenity.
5. The global countryside attracts high levels of non-national property investment, for both commercial and residential purposes.
6. It is not only social and economic relations that are transformed in the global countryside, but also the discursive construction of nature and its management.
7. The landscape of the global countryside is inscribed with the marks of globalization, through deforestation and afforestation; mines and oilfields; tourism infrastructure; the transplantation of plant and animal species; and the proliferation of symbols of global consumer culture, and so on.
8. The global countryside is characterized by increasing social polarization.
9. The global countryside is associated with new sites of political authority.
10. The global countryside is always a contested space.

Massey (2005) calls this the 'throwntogetherness' of place: that place is an ever-shifting constellation of trajectories, enacted through the negotiation of local and global, human and non-human actants. This perspective begins to collapse the notions of local and global, as the global is always entangled in local places. It is through these entanglements that the global can effect changes in local places, but it is also through entanglements in place that the global is produced. As such, the local and the global are seen not in opposition to each other, but in negotiation, with different outcomes in each place producing the uneven geography of globalization. Accordingly, it can be argued that:

> The reconstitution of rural spaces under globalization results from the permeability of rural localities as hybrid assemblages of human and non-human entities, knitted-together intersections of networks and flows that are never wholly fixed or contained at the local scale, and whose constant shape-shifting eludes a singular representation of place. Globalization processes introduce into rural localities new networks of global interconnectivity, which become threaded through and entangled with existing local assemblages, sometimes acting in concert and sometimes pulling local actants in conflicting directions. Through these entanglements, intersections and entrapments, the experience of globalization changes rural places, but it never eradicates the local. Rather, the networks, flows and actors introduced by globalization processes fuse and combine with extant local entities to produce new hybrid formations. In this way, places in the emergent global countryside retain their local distinctiveness, but they are also different to *how they were before*.
>
> (Woods, 2007: 499–500)

This process of geographically variegated place re-constitution through globalization can be illustrated by reference to four brief vignettes. The first vignette, from Youbou on Vancouver Island in western Canada, concerns globalization-linked deindustrialization. In common with many small towns in the Pacific Northwest, Youbou has historically been economically dependent on forestry, and in particular on a sawmill. Forestry, however, has become a globalized industry, with forest products traded internationally and the industry controlled by a handful of transnational corporations, which seek to concentrate production in locations that offer the best cost efficiencies. Small-scale processing plants in Canada and the United States are not favoured by this rationality, and many sawmills and pulpmills have been closed or contracted (Epp and Whitson, 2001). The Cowichan sawmill in Youbou was closed in 2001 by its owners, TimberWest, a Canadian-based transnational corporation, with the loss of 220 jobs. As Prudham (2008) observes, 'paralleling as it did numerous other mill closures in the province over preceding years, the Cowichan mill would have been easy to write off as one more casualty of a fickle global economy' (p. 182). In this case, though, 'the employees of the mill were unwilling to accept the script' (*Ibid*). The former mill workers established the Youbou TimberLess Society (YTS) to campaign for social and environmental justice. When its initial attempts to reverse the mill closure failed, the YTS started to build alliances with environmental groups and with local First Nations communities, eventually concentrating on the setting up of a 'community forest tenure', managed by local people and oriented towards high value-added production (Prudham, 2008). As such, the YTS is at the centre of a re-articulation of place in Youbou, involving the enrolment of human and non-human, local and non-local actants.

The second vignette relates to the negotiation of globalization in Andean mountain communities in South America. As noted earlier in this book, rural areas of South America have long been impacted by global actors and networks, from exploitation by colonial regimes to the development of export-oriented neoliberal economies. Neoliberal-inspired integration into global economic networks presents challenges for the viability of traditional social and economic formations in small rural communities, yet Bebbington (2001) argues that one way in which communities have responded to this pressure has been 'through a progressive engagement in more globalized sets of social and economic relationships, be this via institutional linkages, social relationships, product and labour markets or more generally though an engagement with the modern "development" project' (p. 416). The precise form of these engagements varies between localities, reflecting local resources and capacities and the opportunities for global connections. In Salinas, Ecuador, the response was built on a structure of savings and loan cooperatives set up by an NGO linked to the Catholic Church, which enabled households to buy livestock and to supply community-level cheese factories that were established by the same NGO in partnership with a Swiss dairy programme. In Irupana, Bolivia, fair trade coffee production has been supported by the collaboration of domestic and international NGOs, trade unions, the Catholic Church and a US-funded anti-narcotics coca substitution programme. Much of the product is exported to Belgium, reflecting the involvement of a Belgian NGO in the network. Thus, as Bebbington (2001) comments, 'the global entanglements in which localities are enmeshed are, and have long been multi-stranded: beyond market relationships, the webs linking Andean places and the wider world pass through globalized religious institutions, civil society networks, intergovernmental relationships, migrant streams and more' (p. 415).

The third vignette is from Queenstown, New Zealand. Queenstown is a mountain resort that has long been a popular destination for domestic tourists, but which has over the last 25 years developed into a global amenity resort. Overseas tourists made nearly a million overnight visits in 2007, and contributed NZ$423 million to the local economy in 2004, with many attracted by the resort's branding as 'the adventure capital of the world' (Woods, 2010c). Tourism has been complemented by international amenity migration, fuelling a property development boom and contributing to a doubling of the district's population in 12 years. The transformation has in part been driven by global changes in the transport and tourism industries, investment by global corporations and promotion by global tour operators, but it was also facilitated by New Zealand's neoliberal reforms, notably the liberalization of inward investment controls, privatization of state assets, and the devaluation of the New Zealand dollar, as well as by knock-on effects of agricultural regulation. Moreover, local actors have played a key role as property developers, landowners and business-owners, whilst a radical neoliberal administration at the district council between 1995 and 2001 implemented a laissez-faire approach to new building development, stimulating growth. By 2001, however, local concerns about the rate of development had intensified, including from new in-migrants who feared losing the rural landscape that they had invested in, prompting a fierce debate about the future of the area which was conducted not only locally, but also through national and international media (Woods, 2010c).

The final vignette is about the Larzac plateau in France. A remote, sheep-farming area, the Larzac first came to wider prominence with proposals to establish a military base in the 1980s. Local people were joined in their opposition to the base by anti-militarism campaigners, who came from across France and beyond. When the plans were defeated, many of the campaigners settled in the area, and some took up the local artisan culture of producing Roquefort cheese, including for export. The settlers brought with them their personal histories of travel and involvement in political campaigns at various scales and in various places, as well as new cultural references, which fused with local traditions to produce a distinctive regional counter-culture (Williams, 2008). In 1999, Roquefort cheese was caught in a trade dispute between the United States and the European Union, impacting on the earnings of the farmers. In protest, a group of Larzac farmers belonging to the *Confédération Paysanne* (a radical union of small farmers) and led by Jose Bové attacked and dismantled the McDonald's restaurant under construction in the town of Millau, targeted as a symbol of globalization. In the ensuing trial, Bové called a range of counter-globalization activists as witnesses, utilizing global communications, transport, media and political networks to critique neoliberal globalization. This staged-managed defence was followed by the staging of a larger counter-globalization festival, Larzac 2003. In both cases, the events were made possible by the combining of global networks and a locally embedded counter-culture, such that the Larzac 're-emerged as a focal point in the alterglobalization movement, a place to meet, discuss, organize and to put specific, local concerns in a global context' (Williams, 2008: 52).

In each of these vignettes, rural actors have not been passive recipients of globalization, but have been active agents engaged in contesting, adapting and manipulating globalization forces and networks to help shape their own rural futures. Here, we can think back to questions about the construction of rural space, as discussed in the early chapters of this book. In terms of Halfacree's three-fold model of rural space (see Chapter 1), the proactive engagement of the rural actors in the vignettes demonstrates that the everyday lives of the rural are not overwhelmed or dominated by the broader structural forces that constitute rural space as locality, or by formal representations of the rural. At the same time, the reproduction of rural spaces through the interaction of local and global forces, and local and non-local actors, evidences the relational constitution of the rural (see Chapter 2). Rural places exist as unique meeting-points of diverse social and natural processes and networks, and cannot be understood in isolation from their relationship to other places. Equally, rural places exist as precarious meeting-points of the human and the non-human, as the next section recounts.

RISKY RURALITIES AND THE CO-CONSTITUTED COUNTRYSIDE

Wind, water, fire and pestilence

The story of human presence in the countryside has arguably been the story of the struggle to control and impose meaning on nature. Humans have tamed and cultivated the rural environment for food and clothing; extracted minerals and fuel; captured water flows; modified the landscape and carved out spaces in which to settle and live; and invented spatial orderings that designate wild nature to certain territories set apart from the human (see Chapter 6). Yet, nature has repeatedly failed to conform to these human constructions. Holland and Mooney (2006) describe how early European settlers to New Zealand discovered an environment that was very different to the benign climate and landscape that had been sold to them in promotional materials, and which lay outside their own domestic experiences:

> Settlers who came from rural areas, towns and cities in the British Isles to the tussock grasslands of the eastern South Island of New Zealand encountered physical environmental conditions profoundly different from what they had previously known – broad open landscapes backed by foothills and mountain ranges, large braided rivers prone to flooding at almost any time of year, little natural shelter from wind and rain, and a dearth of wood for fuel, fencing and construction.
>
> (Holland and Mooney, 2006: 39)

As Holland and Mooney record, the settlers adapted by trial and error and by scientifically monitoring the weather and environmental conditions, but they also learned from the indigenous Māori. They learned, for example, that 'Māori associated flowering in kowhai with the onset of spring weather, considered that snow melt in the mountains would lead to flooding in the lowland reach of a large river a day or two later, [and] recognized diurnal variations in the flow rates of large rivers' (*Ibid*: 46). Māori guides also 'taught European travelers how to cross a river in flood, and stressed the importance of rafting a large river in the morning before the wind rose and made navigation difficult in the choppy water' (*Ibid*).

Learning to live with nature has historically been a core part of rural culture, and environmental knowledge has been encoded and passed down through the lay discourses of nature and folklore of indigenous peoples and long-term settler communities. However, in the twentieth century, lay discourses of nature were marginalized by a belief in the transformative capacity of science and technology. When faced with the inhospitable environment of California at the end of the nineteenth century, pioneer resource capitalists manufactured an agricultural Eden through irrigation and drainage schemes, forest and scrubland clearance and soil improvement programmes (see Chapter 3). These same techniques came to be used around the world to expand the spatial reach of commercial agriculture. Similarly, plant and animal species were transported for farming and leisure purposes (see Chapter 2), and biotechnology employed to improve yields and eradicate disease; whilst the construction of bridges and tunnels has opened up new areas for mining, cultivation, settlement and recreation. These scientific and engineering solutions have created temporary fixes in the environment that facilitate the economic exploitation and human settlement of rural areas, so long as nature behaves as anticipated.

The control of water, in particular, has been central to rural settlement and the development of agriculture. Large-scale irrigation schemes in the late nineteenth and early twentieth centuries opened up parts of interior Australia for agriculture, for example, notably in the Murray-Darling Basin, which contains 42 per cent of Australian farmland and accounts for 70 per cent of its irrigation resources. However, farming in marginal arid areas such as interior Australia remains susceptible to drought. At the start of the present century, Australia experienced a major drought that was estimated to cost the national economy at least AU$7 billion, and was popularly represented through 'dramatic pictures of livestock in poor condition, dust storms and barren landscapes' (Alston, 2006: 154). The drought has also had a social impact, which Alston (2006) argues has been experienced differently by gender. For farm men, she suggests, 'drought means a great deal of extra farm work labour carting water and feeding livestock on a daily basis – a grind that after two to five years (depending on the area) was taking its toll on health' (p. 160). Additionally, many farmers 'were finding it difficult to leave their properties, becoming socially isolated because of their workload and weariness' (*Ibid*). The absorption of men in farm work has meant that it is farm women who have had to respond to falling incomes and increasing debt by taking off-farm work, in some cases migrating temporarily to find employment, as well as more generally displaying 'an almost uncritical acceptance of their responsibility for the family's welfare' (p. 159). Women's stories of negotiating and surviving drought, Alston suggests, reveal 'a subjugated discourse about women's experiences of drought that has little to do with the dominant discourse concerning self-reliance, economics and dust storms' (p. 168), and which is therefore neglected in official drought policy. Indeed, since 1990, drought 'has been viewed as a business risk rather than a natural disaster and one to be planned for accordingly' (p. 155; see also, Alston, 2009). This change in discourse, however, has increased the pressure on farm families to show resilience, such that Alston notes how at least one of her interviewees, 'blames herself and her husband for their failure to battle through the drought and for losing their children's inheritance under a mountain of debt' (p. 164).

Controlling water also means managing flood risk, with rural settlements, infrastructure and farmland often concentrated in river valleys and around lakes (Vinet, 2008). Conventionally, flood defence has been constructed as an engineering problem. The Lerma valley in Mexico, home to 2.3 million people, is typical in this respect, with Eakin and Appendini (2008) commenting that the state's 'approach to flooding on the Lerma has been almost exclusively structural, relying on a series of dams, river straightening, and dredging and dike construction' (p. 555). This technocratic model, Eakin and Appendini suggest, has created 'a flood *hazard* from flooding that was previously a well-known and accepted dimension of the hydrology of the Lerma Valley' (p. 556). Historically, rural communities in the valley, as elsewhere in Mexico, manipulated flood events to improve agricultural potential, provide habitats for 'useful' flora and fauna, and expand cultivatable land, as a natural form of irrigation. Annual flooding was an anticipated event, and could be managed through the practices of communal land use. However, attitudes towards flooding have changed with the modernization of agriculture, including the enclosure of private land-holdings and the planting of new crops, such as maize. Additionally, as a consequence of Mexico's entry into NAFTA (see Chapter 8), floodplain land has been released for residential development as peasant smallholdings have become unviable. Flooding has hence been discursively re-articulated as an 'unacceptable risk' (Eakin and Appendini, 2008: 563), and flood defence measures demanded, as a result of land-use changes rather than changes in the behaviour of water. Engineering-based methods can never offer total protection, however, as floods in the Lerma valley in 2003 demonstrated, and Eakin and Appendini note than in several parts of the world, but not yet Mexico, a new approach of 'living with floods' based on returning to traditional methods of managing flood events is being pioneered.

The double edged legacy of technological interventions in nature is also evident with respect to the prevalence of disease in rural societies. Medical advances and improved sanitation, healthcare and environmental management have helped to eradicate or reduce previously endemic diseases in many rural regions. Malaria was eradicated in rural southern Italy, for example, during the first part of the twentieth century in a notable technocratic campaign (Snowden, 2007). At the same time, though, agricultural modernization has been accused of contributing to the emergence of new diseases, with the capacity to transgress human/non-human boundaries and to pass from rural into urban space. Most famously, bovine spongiform encephalopathy (BSE), or 'mad cow disease', was first recorded in cattle in Britain in the 1980s and traced to a mutation of the sheep disease scrapie which had jumped species by means of bonemeal from infected sheep being used in feed for cattle. The acknowledgement in 1996 that BSE could similarly be transmitted to and mutated in humans, and was the most likely source of new variant Creutzfeldt Jakob Disease (vCJD) in humans, not only provoked an immediate collapse in consumer confidence in British beef, the imposition of an export ban on British cattle products, and the implementation of a cull as part of an eradication strategy (Woods, 1998b), but also a wide set of anxieties about the management of risk in the food supply chain (Macnaghten and Urry, 1998; Whatmore, 2002). More recently, the threatened global pandemics in avian flu and swine flu have originated from cross-species transmission in rural areas of South East Asia and Mexico respectively. Although the proximity of humans and animals in traditional farming systems has been blamed for the initial transmission, Davis (2005) accuses technology-driven industrial farming methods of increasing the risk of an epidemic developing:

> A crucial requirement of the modern chicken industry, for example, is 'production density,' the compact location of broiler farms around a large processing plant. As a result, there are now regions in North

> America, Brazil, western Europe, and South Asia with chicken populations in the hundreds of millions – in western Arkansas and northern Georgia, for example, more than 1 billion chickens are slaughtered annually. Similarly, the raising of swine is increasingly centralized in huge operations, often adjacent to poultry farms and migratory bird habitats. The superurbanization of the human population, in other words, has been paralleled by an equally dense urbanization of its meat supply ... Might not one of these artificial Guangdongs be a pandemic crucible as well? Could production density become a synonym for viral density?
>
> (Davis, 2005: 84)

Lay knowledge of nature can extend beyond knowing how to manage natural risks, to knowing how to control and use natural phenomena both practically and in acts of resistance, as can be observed in the case of fire (see also Pyne, 2009). Forest fires and bush fires have become major risks in semi-arid areas of the United States, Australia and Mediterranean Europe, threatening property and human life, and with the potential to encroach on urban landscapes (Davis, 1998). Many fires are started anthropogenically, both accidentally and deliberately, yet fire is also a natural phenomenon that can play a crucial role in the rejuvenation of ecosystems. Moreover, traditional rural communities, including indigenous peoples, have used controlled fires as a way of clearing land for hunting and cultivation or to improve agricultural yields. This has been taken a step further in north-west Spain, where Seijo (2005) argues that 'most individual intentional forest firings taking place in Galicia have a wider meaning as a ritualized act of disaffection with and opposition to the Spanish state's forest policy' (p. 385). The arsonists understand fire as a manageable risk, and see forest fire as 'not being particularly damaging to the peasantry's general welfare while the more disaffected peasants see forest firings as a useful tactic against the state's intrusion in their traditional way of life and a just retaliation against what is perceived as the Spanish state's confiscatory forest policy' (*Ibid*).

Another approach to working with nature has been to turn 'natural disasters' into catalysts for social and economic change. One example is the small town of Greensburg in Kansas, which was hit by a devastating tornado in May 2007, with 11 people killed and many of the town's buildings destroyed. Greensburg had already been suffering from economic and social decline, with a falling population, but the disaster presented an opportunity for renewal. Not only did the town receive nationwide media coverage and considerable donations towards rebuilding, but it was also able to start with a practically blank canvas in planning a new 'model green community' (Harrington, 2010). Many of the new buildings are designed with 'green' features, such as energy conservation, and local businesses have been encouraged to engage with the 'eco-economy' (see Chapter 5). The result, Harrington observes, is that 'some residents who might otherwise have moved away have been energized to remain in Greensburg' (2010: 40).

Climate change and rural futures

The risk from natural hazards in rural areas, as discussed above, is set to increase with the intensification of anthropogenic climate change. Current projections suggest that the disruptive effect of climate change is likely to make extreme weather events such as storms and flooding more frequent, whilst a global rise in temperature will contribute to heightened problems of drought, fire and disease in many regions. More routinely, patterns of agricultural geography will need to adapt to the changed climate. Although some regions will benefit as increased carbon dioxide levels improve yields in crops such as wheat, rice and soybeans, agriculture more broadly will be negatively impacted by higher temperatures, water shortages, reduced soil fertility, more widespread pests, and a higher incidence of heat stress in livestock, altering the appropriate range of different types of farming. Rural tourism will similarly need to adjust to reduced snow cover in traditional mountain resorts and sea level rises threatening coastal resorts. Additionally, the natural landscape and wildlife of rural regions could be modified by climate-linked habitat evolution. In these various ways, non-human agency, in the shape of climate change and its environmental effects, is likely to be the most important factor shaping rural economies, societies and environments over the next century.

One major response is expected to be increased migration, as people leave regions threatened by flooding, affected by food shortages or in which agriculture is no longer viable. This trend can already be observed in sub-Saharan Africa, where rainfall failure has contributed to faster rates of urbanization than observed elsewhere in the global south (Barrios *et al.*, 2006). Africa can also provide several examples of adaptation to recent climate change by rural populations, as Mertz *et al.* (2009) and Osbahr *et al.* (2008) note. In their own work on Senegal, Mertz *et al.* (2009) observe that farmers 'are strongly aware of the climate and have clear opinions on changes, especially in wind patterns and the intensity of climate events' (p. 812), giving rise to adaptations ranging from adopting new crops to youth out-migration. Yet, Mertz *et al.* also report that Senegalese farmers 'have a rather fatalistic approach to climate changes' (p. 814), with a belief that the weather cannot be controlled. Indeed, Osbahr *et al.* (2008) suggest from research in Mozambique that the development of adaptation strategies is complex and multi-scalar, and that many poorer rural residents fall back on traditional methods of 'coping with' rather than 'adapting to' climate change.

Interestingly, far less attention has been devoted to analysing the responsiveness of rural communities in the global north to climate change (but see Harrington, 2005; Hoggart and Henderson, 2005). Here, some of the largest disruptions are likely to come not from the direct impacts of climate change, but from the side-effects of mitigation strategies. A move towards a post-oil society in which private petrol-fuelled transport is limited, for example, would raise considerable questions about the sustainability of commuting, car-based rural tourism and the long-distance transportation of agriculture produce for global markets. For these reasons, conventional wisdom holds that cities offer the most sustainable form of future living. Yet, there is a growing alternative argument that claims that cities are reliant on an oil-based economy and that future populations will need to live and work close to places of renewable energy production, which will be predominantly rural. Some commentators have even suggested that adapting to environmental change will require attributes that are traditionally associated with rurality – solidarity, self-sufficiency and a detailed lay understanding of nature (Farinelli, 2008). This theory has been converted into practice by numerous rural eco-villages, as sites of sustainable rural living (Halfacree, 2007; Meijering *et al.*, 2007). However, these communities tend to be small scale, and it is unclear how a large rural population could really live this way.

Perhaps more probably, rural areas could be enrolled into functions that support sustainable urbanism, through the production of renewable energy and other ecosystem services. For example, both natural and farmed rural landscapes act as significant carbon sinks and carbon locks, including forests, cropland and peat bogs, and the conservation of these landscapes will be important to carbon sequestration (or removing carbon dioxide from the atmosphere), thus mitigating some of the damage of carbon emissions. As discussed briefly in Chapter 8, the ecosystem services approach, based on neoliberal principles, contends that an economic value can be placed on these functions, and that the marketization of these values could make a significant contribution to future rural economies. Gutman (2007) even proposes that ecosystem services could provide the foundation for a 'new rural-urban compact', replacing the historic compact under which rural products and people were sent to the city in return for the city's products, services and governance:

> Some back-of-the-envelope numbers can show us what the economics of such a new rural-urban compact could look like. A world-wide rural ecosystem services bill of [US]$3 trillion a year would be a bargain, considering that estimates of the current value of the world ecosystem are ten to twenty times higher ... Yet $3 trillion a year would more than pay for the annual costs of conservation and the adoption of sustainable agricultural practices worldwide (some [US]$300 billion a year, according to James *et al.*, 1999). It would also be enough to triple the income of the world rural population and still represent no more than 10% of the world GDP.
>
> (Gutman, 2007: 385)

The problem with this argument, in addition to practical challenges in actually implementing payments for ecosystem services in a market economy, is that it conceives of the rural in purely economic and environmental terms. No consideration is given to the cultural or social dimensions of the rural, and in particular to the question of whether rural communities would accept this change in function. Certainly, efforts to promote renewable energy resources in the rural economy have clashed both with agricultural interests and with discourses of the rural idyll. An example of the first case is the controversial expansion of biofuel production, which competes directly with food production and which has been recognized as contributing to increased global food prices (e.g. Saunders *et al.*, 2009). The second case can be illustrated by conflicts over the locating of wind turbine power stations (or 'windfarms') in rural landscapes. Whilst wind turbines are accepted by some rural actors as necessary developments in the long-term interests of the global environment, they can be vehemently opposed by other actors who consider them to be an urban intrusion in the rural landscape, and who mobilize an alternative discourse of nature as constituted by a complex mosaic of localized environments, all of which merit equal protection (Woods, 2003b; Zografos and Martinez-Alier, 2009).

THE RURAL FIGHTS BACK

The politics of the rural

The countryside of the future is likely to be a more contested and politicized space than it has been during the last half century since the end of the Second World War. The dominance of the productivist agricultural discourse of the rural in shaping rural policy during this period, and its wide acceptance among the rural population (and the similar dominance of other primary industry-focused representations in selected rural regions) left little room for contesting the basis of rural policy, at least in the global north. 'Rural politics' during this era was hence largely an arena for debates about territorial management, industrial regulation and the distribution of resources. However, as the material conditions that underpinned this discursive fix began to unravel, conventional 'rural politics' gave way to a new 'politics of the rural', in which the meaning and regulation of rurality itself is the core issue of contestation (Woods, 2003a). The unfolding challenges of globalization and climate change, as discussed above, will further bring into question the function of rural space, and intensify conflicts over its use.

Mormont (1987, 1990) was one of the first to document this transition (although not using these particular terms). Noting that hegemonic representations of rural space had been supplanted by competing representations of rural space that were articulated for the same territory (Mormont, 1990), Mormont proposed that:

> if what could be termed a rural question exists it no longer concerns issues of agriculture or of a particular aspect of living conditions in a rural environment, but questions concerning the specific function of rural space and the type of development to encourage within it.
>
> (Mormont, 1987: 562)

Mormont identified these 'struggles' as emerging first at a local level, over issues such as new building or the proposed closure of key village services such as schools and post offices. Subsequent studies have provided extensive empirical evidence of just such local conflicts developing in rural communities across the global north, especially in communities that have experienced considerable in-migration and population recomposition (see Chapter 6). However, Mormont (1987) also suggested that local campaigns would coalesce over time into new pluri-local movements, collectively taking a stand on issues such as school closures, new house building, windfarm developments, and so on. In this way, rural conflicts can 'jump scales' between the local and the non-local, building a wider 'capacity to act' and enrolling NGOs, politicians and the media into networks that can penetrate the appropriate scale of governance (Cox, 1998). Magnusson and Shaw (2003) similarly argue that the involvement of non-local actors on both sides in a conflict over logging at Clayoquot Sound on Vancouver Island, Canada, reflected a collapsing of 'local' and 'global' scales:

> The politics of places such as Clayoquot puts traditional distinctions between local and global, small and large, domestic and international – and much else – into serious question. If Clayoquot is paradigmatic, it is because the puzzle of politics is especially apparent there.
> (Magnusson and Shaw, 2003: 1)

The significance of sites of conflict such as Clayoquot Sound is that they become emblematic of wider struggles over the meaning and regulation of rural space, with ramifications that will ripple through the countryside. Indeed, some rural conflicts are framed primarily in terms of larger narratives and are driven by external actors, such that local communities become bit-part actors in the story. This can be seen in the shifting terrain of regulation of 'old growth' forests in the United States. For much of the twentieth century, old growth forests were perceived through a discourse of the rural as a resource to be exploited (see Chapter 3), and logged for timber. Towards the end of the century, an alternative discourse started to gain momentum, representing the old growth forests as unique and vulnerable habitats for endangered wildlife such as the spotted owl. As MacDonald (2005) reports:

> The 1990s were heady times for environmentalists. One of their foremost achievements was convincing the average American that there is intrinsic value in 'old-growth' forests. Protecting 'the last best places on earth', a familiar Sierra Club mantra, has become a template

> for environmental activism, a movement fuelled by the controversial albeit wildly successful effort to set aside millions of acres for the northern spotted owl.
>
> (MacDonald, 2005: 2)

Policy fluctuations have created economic and social uncertainty for rural communities in places such as the Allegheny National Forest in Pennsylvania, where logging has traditionally been the focal point for employment, and where planting and regrowth has created a unique forest environment (MacDonald, 2005). Restrictions imposed on logging in the 1990s triggered job losses and depopulation, but in the early 2000s, the Allegheny became the focus for environmental direct action as campaigners claimed that the US Forest Service was conspiring with timber corporations to restart logging and turn the forest into a 'black cherry tree farm' (MacDonald, 2005).

The ascendant neoliberal doctrine that threatened to weaken forest protection has also facilitated a challenge to conservation from the energy sector, particularly in the context of growing international concerns about the imminent watershed of 'peak oil' (the point at which oil production will start to decline due to dwindling reserves) and energy security. The *cause célèbre* here is the Arctic National Wildlife Refuge, an extremely remote and largely pristine wilderness area of 78,050 km^2 in northern Alaska, which also holds oil reserves estimated at between 5.7 and 16 billion barrels. Oil drilling is permissible with Congressional approval, which oil companies have been seeking, supported by a discourse of increasing America's energy security and of boosting the rural economy of Alaska. The opening of the oil fields has been fiercely opposed, however, both by environmental groups and by the indigenous Gwich'in Anthabascan people. As such, Standlea (2006) comments that 'on its face, the Arctic Refuge represents a classic pitched battle between economic growth and environmental conservation' (p. 14), yet he goes on to suggest that 'if we probe the basic, deeper values – worldviews – of the actors and participants in the Arctic Refuge debate, we come up with much more complex and subtle revelations concerning the nature and purpose of development and the very relationship at stake between humans and the nonhuman natural world' (*Ibid*). The issue is not a stark choice between the exploitation of rural resources and the conservation of nature, but rather a far more subtle question about the appropriate balance of different human and non-human components in a hybrid rural. In particular, the use of the refuge area by the Gwich'in Anthabascan for hunting complicates this picture, as Standlea observes:

> The Gwich'in Anthabascan presence in the matrix of actors in the Arctic Refuge truly gives the case a different quality than would be present if this were just about developing oil or saving wildlife and what traditional American environmentalists view as separate 'wilderness', the latter being viewed as some primitive yet uncivilized area still protected from the groping hands of an American unlimited growth worldview depicted in the notion of 'Manifest Destiny'.
>
> (Standlea, 2006: 14)

Interestingly, resistance by indigenous peoples is a feature of several struggles over the exploitation of oil or mineral reserves, including in Manitoba in Canada, Colombia, Ecuador, Guyana, Mexico, Nigeria, the Philippines, and West Papua, with concerns that exploitation would damage both the environmental and the cultural fabric of rural regions (Gedicks, 2001). It is not too great a step to view the cultural interests of non-indigenous but long-standing rural communities in the same way, especially in cases where there is conflict between mining and farming, as in parts of Australia.

The perceived marginalization of local rural communities in conflicts over the use of rural space has contributed to an increasing sense of beleaguerment and neglect by rural communities in many parts of the global north. Together with discontent about declining support for family farming, the rationalization of rural services, restrictions on the development of rural settlements, and challenges to traditional rural pursuits such as hunting, this perception has given rise to a new 'rural identity movement' (Woods, 2003a, 2008c). This movement has antecedents in earlier farmers' movements in several countries (e.g. Halpin, 2004; Stock, 1996), and in the 'rural community movement' in northern Scandinavia (Halhead, 2006), but is distinctive in its articulation of a rural–urban divide as a framing device (Woods, 2008c).

One of the most prominent examples is the Countryside Alliance and associated groups in Britain. Established to front opposition to the proposed ban on the hunting of wild mammals with hounds in England and Wales (see Chapter 7), the Countryside Alliance explicitly framed its campaign in terms of an assault on rural Britain by an urban elite, positioning hunting as central to the rural way of life and also extending its campaigning to other issues such as farming and rural service provision (Woods, 2005b). Moreover, the Countryside Alliance has formed the nucleus of a diffuse and disconnected assemblage of rural protest groups, including ad hoc local campaigns, more specialist organizations such as Farmers for Action, and militant direct action groups such as the Countryside Action Network and the 'Real Countryside Alliance' (Woods, 2005b). Internationally, the mobilization of the rural identity movement has been similarly diffuse, incorporating mass protests and demonstrations (Britain, France, Belgium), new campaign and lobbying groups (Australia, Britain, Ireland, United States), rural political parties and maverick independent candidates (Australia, France, New Zealand), militant direct action (Britain, France, Spain), and right-wing militia groups (United States) (Woods, 2008c).

Reclaiming the land

A different trajectory has simultaneously fired the mobilization of new rural social movements in the global south, which share some of the characteristics and concerns of the rural identity movement in the global north, but which are founded in the particular history of colonialism, democratization and under-development (Woods, 2008c). Modern rural social movements emerged in many developing countries during the 1960s and 1970s, influenced both by Marxism and by Liberation Theology, with protests directed against externally driven modernization programmes (see Chapter 5) and repressive state action. These mobilizations faded with de-peasantization, leading some commentators to write off the political potential of the countryside (Bernstein, 2002; de Janvry, 1981). Others, however, have identified the resurgence of a 'new wave' of social movements in the global south, characterized by one or more of a mixed social base of rural peasants and urban proletarians, leadership from 'peasant intellectuals', an 'anti-political' strategy and use of direct action tactics, an internationalist vision and the ideological fusing of Marxism and ethnic politics (Moyo and Yeros, 2005; Petras, 1997).

Rural social movements in the south have tended to focus on two key issues. First, land rights have been the major concern for rural social movements in Latin America and southern Africa (Moyo and Yeros, 2005).

In some cases, the emphasis has been on the redistribution of land in rural areas, challenging the colonial legacy of large estates and corporate land-holdings, and in other cases on resettling landless rural migrants or on resisting the sequestration of property from peasant farmers (Woods, 2008c). Many of these movements have combined political protests with actions for social transformation, including land occupations (Moyo and Yeros, 2005; Wolford, 2004). Second, peasant farmers' movements have mobilized to defend traditional forms of agriculture against neoliberal reforms and the practices of transnational agri-food corporations, especially in India and South East Asia. The successful opposition of the Karnataka State Farmers' Union in India to the promotion of genetically modified cotton seeds and field trials by Cargill and Monsanto, is often held-up as a prime example (Routledge, 2003).

In many cases, issues of land reform and support for peasant agriculture are necessarily entwined, and are further complemented by concerns with wider aspects of social and economic conditions in rural areas, including campaigning for improvements in electrification, sanitation and education (Bentall and Corbridge, 1996). As such, social movements in the global south advocate a discourse of rurality that is at the same time both grounded in tradition – in the veneration of the peasant way of life – and socially progressive. Furthermore, many groups are actively engaged in working towards these goals materially, including the *Movimento dos Trabalhadores Rurais Sem-Terra* (MST) in Brazil. The MST, or landless workers movement, has established a number of compounds on which families have been resettled and given land to work, in addition to its engagement in political protests and lobbying (Wolford, 2004). In the 20 years following its formation in 1984, the MST resettled over 350,000 families on 2,200 compounds (Wittman, 2009). The compounds are used as a vehicle for providing education and health promotion, and are managed in ways that promote afforestation and the practice of sustainable agriculture. In this way the compounds function as a test-bed for the MST's rural vision, what Wittman (2009) calls a 'new agrarian citizenship'. Yet, as Caldeira (2008) shows, the enthusiasm of the MST leadership for this agenda is not always shared by grassroots participants, who are preoccupied with more basic material concerns.

Rural social movements in the south are spatially transgressive in two ways. First, they can transgress the rural–urban divide by working to organize ex-rural migrants in urban areas. Second, the movements also transgress national boundaries, forming transnational alliances such as Peoples' Global Action and *Via Campesina* (Borras *et al.*, 2008; Desmarais, 2008; Featherstone, 2008; Routledge and Cumbers, 2009). They also recognize the need to transport their concerns into northern (and urban) public spaces in order to be heard. Featherstone (2003) hence describes the 'inter-continental caravan' of Indian farmers through Europe to promote their concerns, whilst international farmers' days of action have become a regular feature at meetings of the G20 and the WTO (Featherstone, 2008; Routledge and Cumbers, 2009). In presenting this alternative discourse outside the compound of the major meeting, the social movements have become active participants in the re-articulation of the rural. However, the rural vision presented by the protesters is not always coherent and internal tensions exist within alliances such as Via Campesina over the precise articulation of rural interests and characteristics (Desmarais, 2008). Routledge (2003) accordingly describes events such as demonstrations outside summits as 'convergence spaces' which provide a temporary discursive fix to enable diverse interests to come together around a key set of demands or objectives. Tellingly, though, the recruitment of farm unions from the global north to Via Campesina remains limited, reflecting the difficulty of persuading farmers in Australia, Europe or North America that they have shared interests with farmers in India or Peru.

CONCLUSION: A RELATIONAL RURAL GEOGRAPHY

At some point during the opening decade of the twenty-first century, the global urban population surpassed the global rural population for the first time in history. Many countries in Europe and North America have had an urban majority population for around a century: Britain crossed the threshold in the 1850s, Germany around 1900, France and the United States by 1920, Canada in 1931. The global tipping point has come with rapid urbanization in Brazil, China and India, and other fast-growing countries of the global south. Yet, the population shift does not in itself necessarily mean that the rural has been eclipsed, or become irrelevant. On the contrary, as this book has demonstrated, the rural continues to be central to many of the key issues confronting the world today, and the study of rural geographies is arguably as important as ever.

However, the demands placed on rural space by concerns such as food security and protecting biodiversity are not necessarily complementary, and in many cases they challenge or disrupt the settled discursive and material geographies of people living in rural communities. For example, as this chapter has discussed, rural areas will not only be directly affected by climate change, but also potentially have an important role to play in building a more sustainable society as a source of renewable energy and ecosystem services. Yet, the development of these new functions may involve changes to the rural landscape or to established rural practices that will offend certain discourses of rurality, as has already been demonstrated in opposition to renewable energy projects such as windfarms.

Rural communities are not the passive recipients and victims of changes imposed from outside, but have the capacity to mobilize to represent their own interests. Indeed, the perceived marginalization of rural interests in an increasingly urbanized world has prompted the mobilization of new rural social movements that have become a feature of contemporary politics in both the global north and the global south. The coalescence of these groups into transnational alliances has further provided a focal point for resistance to neoliberal globalization, taking the struggle over the meaning and regulation of rurality to the global stage.

The enduring significance of the rural therefore lies in its relationality. The rural is not a pre-determined and discrete geographical territory, and neither is it a fantasy of the imagination. Rather, viewed from a relational perspective, the rural comprises millions of dynamic meeting-points, where different networks, and flows and processes are knotted together in unique ways (see Chapter 2). These configurations are enacted through the everyday lives of rural people (and, indeed, the lives of non-human rural residents), and they are given meaning by the application of particular ideas of rurality. The 'family farm', for instance, is an entanglement of social and economic processes, labour and family relations, cultural conventions and landscape practices, that has material form, performed expression and discursive symbolism as an icon of rurality.

By implication, we can see that rural change occurs by modifying the individual components in rural configurations, substituting them for different components, or rearranging existing components in new ways. Productivism, for example, modified the components that configured the family farm, transforming them into something that might retain elements of the family farm, but which is now materially very different to the discursive ideal (see Chapter 3). More broadly, rural restructuring has affected substitutions that have brought rural configurations closer in form and character to the configurations that constitute urban space and society, yet this process has not replicated urban forms within the countryside, but has given rise to new hybrid articulations of the twenty-first century rural.

The dynamism of the contemporary rural has inevitably produced tensions and conflicts that are expressions of the relational politics of the rural. Following Amin (2004), we can identify both a *politics of propinquity*, in which the juxtaposition of competing demands on the rural produces friction (for example, between ideas of environmental conservation and respect for traditional rural cultures), and a *politics of connectivity*, in which the integration of rural places into wider social, economic and political systems exposes them to the effects of distant events and decisions (for example, the impact of a trade agreement on the viability of an individual farm). Investigating these relational politics is a key challenge for rural geographers and has implications for the way in which rural geography is practised.

Exploring the relationality of the rural requires examination of each of the three points in Halfacree's (2006) three-fold model of rural space (see Chapter 1). The portal of 'rural locality' allows us to glimpse the structural patterns produced by configurations of larger social and economic processes; the portal of 'representations of the rural' provides sight of the discursive meanings applied to the rural in relation to the wider world; and the portal of the 'everyday lives of the rural' illuminates the routine enactment of a relational rural by individuals whose mobility is not constrained to rural space. These perspectives each draw on different conceptual and methodological tools within the rural geographer's toolbox. Political-economy analysis remains important for understanding the structuring of rural localities; the theory of social construction and techniques of discourse analysis enable study of representations of the rural; and the developing body of performance research in rural geography engages with the everyday lives of the rural. As such, the trajectory of conceptual and methodological developments in rural geography over the last 30 years can be regarded as an accumulation of capacity, rather than as a series of sharp changes in direction.

At the same time, a relational rural geography will expand the boundaries of rural research and lead rural geographers into new associations.

Recognition of the global inter-connection and inter-dependency of rural places points to a dismantling of the separation between rural research on the global north and rural research on the global south, and the promotion of more transnational research. Similarly, there is scope for collaboration between rural geographers and urban geographers, and for studies combining rural and urban research, in teasing out the messy entanglements of the rural and the urban. Finally, as rural places are composed of both human and non-human actants, and are subject to both human and non-human agency, new insights may be gained through engaging with natural and physical scientists in truly inter-disciplinary research (see Lowe and Phillipson, 2006).

The rural is, and always has been, a dynamic and diverse space, made elusive by its relationality. The idea of the rural has had a powerful resonance throughout history, and has attracted, inspired and confounded geographers in equal measure. As Murdoch (2003) noted, there is no single vantage point from which the whole of the rural can be observed. Our studies give only partial glimpses. Yet, it is this complexity of the rural that makes the study of rural geography challenging and exciting, and we have much still to explore.

FURTHER READING

Further discussion of globalization and the reconstitution of rural places within an emergent 'global countryside' can be found in my paper, 'Engaging the Global Countryside' published in *Progress in Human Geography* (Woods, 2007). As noted, there are relatively few locality-based studies of the impact of globalization on rural communities, but some examples include Anthony Bebbington's discussion of the 'globalized Andes' in *Ecumene* (now *Cultural Geographies*) in 2001, Echánove's analysis of a Mexican rural community in *Tidjschrift voor Economische en Sociale Geografie* (2005), and my article on Queenstown, New Zealand, in *Geojournal* (Woods, 2010c). The adaptation of European settlers to environmental conditions in rural New Zealand is discussed by Peter Holland and Bill Mooney in *New Zealand Geographer* (2006). Other dimensions of the risky co-constitutiveness of the rural are explored further by Margaret Alston on gendered experiences of drought in Australia (*Sociologia Ruralis*, 2006) and by Hallie Eakin and Kirsten Appendini on managing flood risk in Mexico (*Agriculture and Human Values*, 2008). Analyses of rural livelihood adaptation to climate change have been conducted in Senegal by Ole Mertz *et al.* (*Environmental Management*, 2009) and in Mozambique by Henny Osbahr *et al.* (*Geoforum*, 2008). Pablo Gutman's argument that ecosystem services can form the basis of a new rural–urban compact can be found in *Ecological Economics* (2007). A special issue of the *Journal of Rural Studies* in 2008 (volume 24, issue 2) provides further discussion of rural social movements, including an overview introduction (Woods, 2008c), an examination of Via Campesina by Annette Desmarais, and Scott Prudham's account of the community response to globalist forestry in Youbou, Canada. Good collections on rural social movements in the global south have been edited by Borras, Edelman and Kay (*Transnational Agrarian Movements*, 2008) and by Moyo and Yeros (*Reclaiming the Land*, 2005). Hannah Wittman's paper in the *Journal of Rural Studies* (2009) and Wendy Wolford's article in the *Annals of the Association of American Geographers* (2004) provide more detailed studies of the landless workers movement (MST) in Brazil.

Bibliography

Abram, S. (2003) The rural gaze, in Cloke, P. (ed.) *Country Visions*. Harlow: Pearson.

Addley, E. (2008) Welcome to Thanet Earth: Is this a taste of the future for UK agriculture? *The Guardian*, 11 June.

Agarwal, S., Rahman, S. and Errington, A. (2009) Measuring the determinants of relative economic performance in rural areas, *Journal of Rural Studies*, 25: 309–21.

Agyeman, J. and Spooner, R. (1997) Ethnicity and the rural environment, in Cloke, P. and Little, J. (eds) *Contested Countryside Cultures*. London: Routledge.

Alder, J. (1989) Origins of sightseeing, *Annals of Tourism Research*, 16: 7–29.

Allen, P., FitzSimmons, M., Goodman, M. and Warner, K. (2003) Shifting plates in the agrifood landscape: the tectonics of alternative agrifood initatives in California, *Journal of Rural Studies*, 19: 61–76.

Almagor, U. (1985) A tourist's 'vision quest' in an African game reserve, *Annals of Tourism Research*, 12: 31–47.

Alphandéry, P. and Fortier, A. (2001) Can a territorial policy be based on science alone? The system for creating the Natura 2000 network in France, *Sociologia Ruralis*, 41: 311–28.

Alston, M. (2006) 'I'd like to just walk out of here': Australain women's experience of drought, *Sociologia Ruralis*, 46: 154–70.

Alston, M. (2009) Drought policy in Australia: gender mainstreaming or gender blindness? *Gender, Place and Culture*, 16: 139–54.

Amin, A. (2004) Regions unbound: towards a new politics of place, *Geografiska Annaler*, 86B: 33–44.

Anderson, C. and Bell, M. (2003) The devil of social capital: a dilemma of American rural sociology, in Cloke, P. (ed.) *Country Visions*. Harlow: Pearson.

Anderson, K. and Valenzuela, E. (2008) The softest subsidy: agricultural subsidy cuts, new biotechnologies, developing countries, and cotton, *Georgetown Journal of International Affairs*, 9(1): 7–16.

Angeles, L.C. and Hill, K. (2009) The gender dimension of the agrarian transition: women, men and livelihood diversification in two peri-urban farming communities in the Philippines, *Gender, Place and Culture*, 16: 609–29.

Appadurai, A. (1996) *Modernity at Large: Cultural dimensions of globalization*. Minneapolis: University of Minnesota Press.

Argent, N. (2002) From pillar to post? In search of the post-productivist countryside in Australia, *Australian Geographer*, 33: 97–114.

Armesto López, X.A. and Gómez Martín, B. (2006) Tourism and quality agro-food products: an opportunity for the Spanish countryside, *Tijdschrift voor Economische en Sociale Geografie*, 97: 166–77.

Árnason, A., Shucksmith, M. and Vergunst, J. (eds) (2009) *Comparing Rural Development*. Aldershot: Ashgate.

Ashley, P. (2007) Toward an understanding and definition of wilderness spirituality, *Australian Geographer*, 38: 53–69.

Atkins, P. and Bowler, I. (2001) *Food in Society: Economy, culture, geography*. London: Arnold.

Ayto, J. (1990) *Dictionary of Word Origins*. London: Bloomsbury.

Baker, K. and Jewitt, S. (2007) Evaluating thirty-five years of Green Revolution technology in villages of Bulandshahr District, Western UP, North India, *Journal of Development Studies*, 43: 312–29.

Baker, S. and Brown, B.J. (2008) Habitus and homeland: educational aspirations, family life and culture in autobiographical narratives of educational experience in rural Wales, *Sociologia Ruralis*, 48: 57–72.

Barnett, A. (1998) Securing the future, in A. Barnett and R. Scruton (eds) *Town and Country*. London: Jonathan Cape.

Barrett, H., Ilbery, B., Browne, A. and Binns, T. (1999) Globalization and the changing networks of food supply: the importation of fresh horticultural produce from Kenya into the UK, *Transactions of the Institute of British Geographers*, 24: 159–74.

Barrios, S., Bertinelli, L. and Strobl, E. (2006) Climatic change and rural-urban migration: the case of sub-Saharan Africa, *Journal of Urban Economies*, 60: 357–71.

Bebbington, A. (1999) Capitals and capabilities: a framework for analyzing peasant viability, rural livelihoods and poverty, *World Development*, 27: 2021–44.

Bebbington, A. (2001) Globalized Andes? Livelihoods, landscapes and development, *Ecumene*, 8: 414–36.

Bell, C. (1997) The 'real' New Zealand: rural mythologies perpetuated and commodified, *Social Science Journal*, 34: 145–58.

Bell, D. (1997) Anti-idyll: rural horror, in Cloke, P. and Little, J. (eds) *Contested Countryside Cultures: Otherness, Marginalisation and Rurality*. London: Routledge.

Bell, D. (2006) Variations on the rural idyll, in Cloke, P., Marsden, T. and Mooney, P. (eds) *Handbook of Rural Studies*. London: Sage.

Bell, M. (1994) *Childerley: Nature and Morality in a Country Village*. Chicago: University of Chicago Press.

Bell, M. (2007) The two-ness of rural life and the ends of rural scholarship, *Journal of Rural Studies*, 23: 402–15.

Bentall, J. and Corbridge, S. (1996) Urban-rural relations, demand politics and the 'new agrarianism' in northwest India: the Bharatiya Kisan Union, *Transactions of the Institute of British Geographers*, 21: 27–48.

Berger, J. (1972) *Ways of Seeing*, Harmondsworth: Penguin.

Bernard, T., Collion, M.-H., de Janvry, A., Rondot, P. and Sadoulet, E. (2008) Do village organizations make a difference in African rural development? A study for Senegal and Burkina Faso, *World Development*, 36: 2188–204.

Bernstein, H. (2002) Land reform: taking a long(er) view, *Journal of Agrarian Change*, 2: 433–63.

Berry, B. (ed.) (1976) *Urbanization and Counterurbanization*. Beverley Hills: Sage.

Berry, W. (2009) *Home Economics: Fourteen Essays*. Berkeley, CA: Counterpoint.

Bertrand, N. and Kreibich, V. (eds) (2006) *Europe's City-Regions Competitiveness: Growth, regulation and peri-urban land management*. Assen, NL: Royal Van Gorcum.

Besio, K., Johnston, L. and Longhurst, R. (2008) Sexy beasts and devoted mums: narrating nature through dolphin tourism, *Environment and Planning A*, 40: 1219–34.

Best, S. (1989) The commodification of reality and the reality of commodification: Jean Baudrillard and post-modernism, *Current Perspectives in Social Theory*, 19: 23–51.

Bibby, P. and Shepherd, J. (2004) *Developing a New Classification of Urban and Rural Areas for Policy Purposes – Methodology*. Rural Evidence Research Centre Working Paper. London: RERC. Available http: <http://www.statistics.gov.uk/geography/downloads/Methodology_Reports.pdf>

Binswanger, H.P. (2007) Empowering rural people for their own development, *Agricultural Economics*, 37: 13–27.

Bjarnason, T. and Thorlindsson, T. (2006) Should I stay or should I go? Migration expectations among youths in Icelandic fishing and farming communities, *Journal of Rural Studies*, 22: 290–300.

Bjørkhaug, H. and Richards, C.A. (2008) Multifunctional agriculture in policy and practice? A comparative analysis of Norway and Australia, *Journal of Rural Studies*, 24: 98–111.

Bone, R.M. (2003) *The Geography of the Canadian North*. Don Mills, ON: Oxford University Press.

Borras, S.M., Edelman, M. and Kay, C. (eds) (2008) *Transnational Agrarian Movements: Confronting Globalization*. Chichester: Wiley-Blackwell.

Bowen, S. and Valenzuela Zapata, A. (2009) Geographical indications, *terroir*, and socioeconomic and ecological sustainability: the case of tequila, *Journal of Rural Studies*, 25: 108–19.

Bowler, I. (1985) Some consequences of the industrialisation of agriculture in the European Community, in Healey, M. and Ilbery, B.W. (eds) *The Industrialization of the Countryside*. Norwich: Geo Books.

Boyle, P. and Halfacree, K. (1998) *Migration into Rural Areas*. Chichester: Wiley.

Brace, C. (2003) Rural mappings, in Cloke, P. (ed.) *Country Visions*. Harlow: Pearson.

Brandth, B. (2002) Gender identity in European family farming: a literature review, *Sociologia Ruralis*, 42: 181–200.

Brandth, B. (2006) Agricultural body-building: incorporations of gender, body and work, *Journal of Rural Studies*, 22: 17–28.

Brennan, M.A., Flint, C.G. and Luloff, A.E. (2009) Bringing together local culture and rural development: findings from Ireland, Pennsylvania and Alaska, *Sociologia Ruralis*, 49: 97–112.

Brennan-Horley, C., Connell, J. and Gibson, C. (2007) The Parkes Elvis Revival Festival: Economic development and contested place identities in rural Australia, *Geographical Research*, 45: 71–84.

Bressey, C. (2009) Cultural archaeology and historical geographies of the black presence in rural England, *Journal of Rural Studies*, 25: 386–95.

Brockington, D. and Igoe, J. (2006) Eviction for Conservation: a global overiew, *Conservation and Society*, 4: 424–70.

Browne, W.P. (2001) *The Failure of National Rural Policy*. Washington, DC: Georgetown University Press.

Bruce, D. and Whitla, M. (eds) (1993) *Community-Based Approaches to Rural Development*. Sackville, NB: Mount Allison University.

Bruinsma, J. (ed.) (2003) *World Agriculture: towards 2015/2030 – an FAO perspective*. London: Earthscan.

Buck, D., Getz, C. and Guthman, J. (1997) From farm to table: the organic vegetable commodity chain of northern California, *Sociologia Ruralis*, 37: 3–20.

Buller, H. (2004) Where the wild things are: the evolving iconography of rural fauna, *Journal of Rural Studies*, 20: 131–42.

Bunce, M. (1994) *The Countryside Ideal: Anglo-American Images of Landscape*, London and New York: Routledge.

Bunce, M. (2003) Reproducing rural idylls, in Cloke, P. (ed.) *Country Visions*. Harlow: Pearson.

Burnley, I. and Murphy, P. (2004) *Sea Change: Movement from Metropolitan to Arcadian Australia*. Sydney: UNSW Press.

Burton, R.J.F. (2004) Seeing through the 'good farmer's' eyes: towards developing an understanding of the social symbolic value of 'productivist' behaviour, *Sociologia Ruralis*, 44: 195–215.

Busch, L. and Bain, C. (2004) New! Improved? The transformation of the global agrifood system, *Rural Sociology*, 69: 321–46.

Buttel, F. and Newby, H. (eds) (1980) *The Rural Sociology of Advanced Societies: Critical Perspectives*. Montclair, NJ: Allanheld and London: Croom Held.

Bye, L.M. (2003) Masculinity and rurality at play in stories about hunting, *Norsk Geografisk Tidsskrift – Norwegian Journal of Geography*, 57: 145–53.

Bye, L.M. (2009) 'How to be a rural man': young men's performances and negotiations of rural masculinities, *Journal of Rural Studies*, 25: 278–88.

Caldeira, R. (2008) My land, your social transformation: conflict within the Landless People Movement (MST), Rio de Janeiro, Brazil, *Journal of Rural Studies*, 24: 150–60.

Campbell, H. (2000) The glass phallus: pub(lic) masculinity and drinking in rural New Zealand, *Rural Sociology*, 65: 562–81.

Carolan, M.S. (2008) More-than-representational knowledge/s of the countryside: how we think as bodies, *Sociologia Ruralis*, 48: 408–22.

Carroll, M.S. (1995) *Community and the Northwestern Logger*. Boulder, CO: Westview.

Carson, R. (1962) *Silent Spring*. Cambridge, MA: Riverside Press.

Casid, J.H. (2005) *Sowing Empire: Landscape and colonization*. Minneapolis: University of Minnesota Press.

Castree, N. (2008a) Neoliberalising nature: the logics of deregulation and reregulation, *Environment and Planning A*, 40: 131–52.

Castree, N. (2008b) Neoliberalising nature: processes, effects, and evaluations, *Environment and Planning A*, 40: 153–73.

Cater, C. and Smith, L. (2003) New country visions: adventurous bodies in rural tourism, in Cloke, P. (ed.) *Country Visions*. Harlow: Pearson.

Catlin, G. (1930) *Letters and Notes on the Manners, Customs and Conditions of North American Indians*, vol. 1. Edinburgh: John Grant.

Cawley, M. and Gillmor, D. (2008) 'Culture economy', 'integrated tourism' and 'sustainable rural development': evidence from Western Ireland, in Robinson, G.M. (ed.) *Sustainable Rural Systems*. Aldershot: Ashgate.

Chakraborti, N. and Garland, J. (eds) (2004) *Rural Racism*. Cullompton, UK: Willan.

Chambers, R. (1983) *Rural Development: Putting the Last First*. London: Longman.

Chambers, R. (1993) *Challenging the Professions: Frontiers for Rural Development*. London: Intermediate Technology Publications.

Chambers, R. (1994) The origins and practice of participatory rural appraisal, *World Development*, 22: 953–69.

Chambers, R. and Conway, G. (1992) Sustainable rural livelihoods: practical concepts for the twenty-first century, *IDS Discussion Paper 296*, Brighton: Institute for Development Studies.

Chaverri, P.P. (2006) Cultural and environmental amenities in peri-urban change: the case of San Antonio de Escazú, Costa Rica, in Moss, L.A.G. (ed.) *The Amenity Migrants: Seeking and sustaining mountains and their cultures*. Wallingford: CABI.

Cheshire, L. (2006) *Governing Rural Development*. Aldershot: Ashgate.

Cheshire, L. and Woods, M. (2009) Rural citizenship and governmentality, in Kitchin, R. and Thrift, N. (eds) *International Encyclopaedia of Human Geography*. Oxford: Elsevier.

Ching, B. and Creed, E.W. (eds) (1997) *Knowing Your Place: Rural identity and cultural hierarchy*. New York: Routledge.

Clark, G. (1991) People working in farming: the changing nature of farmwork, in Champion, T. and Watkins, C. (eds) *People in the Countryside*, London: Paul Chapman.

Clemenson, H. (1992) Are single industry towns diversifying? An examination of fishing, forestry and mining towns, in Bollmann, R.D. (ed.) *Rural and Small Town Canada*. Toronto: Thompson Educational Publishing.

Cloke, P. (1977) An index of rurality for England and Wales, *Regional Studies*, 11: 31–46.

Cloke, P. (1989a) Rural geography and political economy, in Peet, R. and Thrift, N. (eds) *New Models in Geography: The Political Economy Perspective*, vol. 1. London: Unwin Hyman.

Cloke, P. (1989b) State deregulation and New Zealand's agricultural sector, *Sociologia Ruralis*, 29: 34–47.

Cloke, P. (1993) The countryside as commodity: new rural spaces for leisure, in Glyptis, S. (ed.) *Leisure and the Environment*. London: Bellhaven.

Cloke, P. (1994) (En)culturing political economy: a life in the day of a 'rural geographer', in Cloke, P., Doel, M., Matless, D., Phillips, M. and Thrift, N. (eds) *Writing The Rural*. London: Paul Chapman.

Cloke, P. (2004) Rurality and racialised others: out of place in the countryside?, in Chakraborti, R. and Garland, J. (eds) *Rural Racism*. Cullompton, UK: Willan.

Cloke, P. (2006) Conceptualizing rurality, in Cloke, P., Marsden, T., and Mooney, P. (eds) *Handbook of Rural Studies*. London: Sage.

Cloke, P. and Edwards, G. (1986) Rurality in England and Wales 1981: a replication of the 1971 index, *Journal of Rural Studies*, 20: 289–306.

Cloke, P. and Jones, O. (2001) Dwelling, place and landscape: an orchard in Somerset, *Environment and Planning A*, 33: 649–66.

Cloke, P. and Jones, O. (2002) *Tree Cultures*. Oxford: Berg.

Cloke, P. and Little, J. (eds) (1997) *Contested Countryside Cultures*. London: Routledge.

Cloke, P. and Perkins, H.C. (1998) Cracking the canyon with the awesome foursome: representations of adventure tourism in New Zealand, *Environment and Planning D: Society and Space*, 16: 185–218.

Cloke, P. and Perkins, H.C. (2002) Commodification and adventure in New Zealand tourism, *Current Issues in Tourism*, 5: 521–49.

Cloke, P. and Perkins, H.C. (2005) Cetacean performance and tourism in Kaikoura, New Zealand, *Environment and Planning D: Society and Space*, 23: 903–24.

Cloke, P., Phillips, M. and Thrift, N. (1995) The new middle classes and the social constructs of rural living, in Butler, T. and Savage, M. (eds) *Social Change and the Middle Classes*. London: UCL Press.

Cloke, P., Phillips, M. and Thrift, N. (1998) Class, colonisation and lifestyle strategies in Gower, in Boyle P. and Halfacree, K. (eds) *Migration to Rural Areas*, London: Wiley.

Clout, H.D. (1972) *Rural Geography: an introductory survey*. Oxford: Pergamon Press.

Coates, K. (2001) Northland: The past, present and future of northern British Columbia in an age of globalization, in Epp, R. and Whitson, D. (eds) *Writing Off the Rural West*. Edmonton: University of Alberta Press.

Cockburn, A. (1996) A short, meat-oriented history of the world: from Eden to the Mattole, *New Left Review*, 215: 16–42.

Cocklin, C. and Dibden, J. (2002) Taking stock: Farmer's reflections on the deregulation of Australian dairying, *Australian Geographer*, 33: 29–42.

Cohen, A. (1985) *The Symbolic Construction of Community*. London: Tavistock.
Coldwell, I. (2007) New farming masculinities: 'More than just shit-kickers', we're 'switched-on' farmers wanting to 'balance lifestyle, sustainability and coin', *Journal of Sociology*, 43: 87–103.
Connell, J. and Gibson, C. (2003) *Sound Tracks: Popular music, identity and place*. London: Routledge.
Connors, T. (1996) *To Speak with One Voice: The quest by Australian farmers for federal unity*. Canberra: National Farmers Federation.
Convery, I., Bailey, C., Mort, M. and Baxter, J. (2005) Death in the wrong place? Emotional geographies of the UK 2001 foot and mouth disease epidemic, *Journal of Rural Studies*, 21: 98–109.
Corbett, M. (2007a) *Learning to Leave: The irony of schooling in a coastal community*. Halifax, NS: Fernwood.
Corbett, M. (2007b) All kinds of potential: women and out-migration in an Atlantic Canadian coastal community, *Journal of Rural Studies*, 23: 430–42.
Cosgrove, D. (1985) Prospect, perspective and the evolution of the landscape idea, *Transactions of the Institute of British Geographers*, 10: 45–62.
Courtney, P., Short, C., Kambites, C., Moseley, M., Ilbery, B., Boase, R., Owen, S. and Clark, M. (2007) *The Social Contribution of Land-based Industries to Rural Communities*. Cheltenham: Commission for Rural Communities.
Cox, G., Hallett, J. and Winter, M. (1994) Hunting the wild red deer: the social organization and ritual of a 'rural' institution, *Sociologia Ruralis*, 34: 190–205.
Cox, K.R. (1998) Spaces of dependence, spaces of engagement and the politics of scale, or: looking for local politics, *Political Geography*, 17: 1–24.
CPRE (Campaign to Protect Rural England) website: www.cpre.org.uk (accessed 6 June 2009).
Crang, M. (1999) Nation, region and homeland: history and territory in Darlana, Sweden, *Ecumene*, 6: 447–70.
CRC (2007) *State of the Countryside 2007*. Cheltenham, UK: Commission for Rural Communities.
Cronon, W. (1996) The trouble with wilderness, *Environmental History*, 1: 7–28.
Crouch, D. (2006) Tourism, consumption and rurality, in Cloke, P., Marsden, T. and Mooney, P. (eds) *Handbook of Rural Studies*. London: Sage.
Daniels, S. (1989) Marxism, culture and the duplicity of landscape, in Peet, R. and Thrift, N. (eds) *New Models in Geography*, vol. 2, London: Unwin Hyman.
Daugstad, K. (2008) Negotiating landscape in rural tourism, *Annals of Tourism Research*, 35: 402–26.
David, P. and Wright, G. (1997) Increasing returns and the genesis of American resource abundance, *Industrial and Corporate Change*, 6: 203–45.
Davis, M. (1998) *Ecology of Fear: Los Angeles and the imagination of disaster*. New York: Metropolitan Books.
Davis, M. (2005) *The Monster at our Door: The global threat of avian flu*, New York: New Press.

Dean, M. (1999) *Governmentality: Power and rule in modern society*. London: Sage.

Defra (2002) *Public Attitudes to Quality of Life and the Environment*. London: Department for Environment, Food and Rural Affairs.

Defra (2007) *Rural Development Plan for England*. London: Department for Environment, Food and Rural Affairs.

Defra (2009) Organic Statistics Dataset. London: Department for Environment, Food and Rural Affairs. Available HTTP: <http://www.defra.gov.uk/evidence/statistics/foodfarm/enviro/organics/documents/organics-2009.xls

De Janvry, A. (1981) *The Agrarian Question and Reformism in Latin America*. Baltimore: Johns Hopkins University Press.

Demeritt, D. (2001) Scientific forest conservation and the statistical picturing of nature's limits in the Progressive-era United States, *Environment and Planning D: Society and Space*, 19: 431–59.

Derkzen, P. (2010) Rural partnerships in Europe, a differentiated view from a country perspective: the Netherlands and Wales, *European Urban and Regional Studies*, 17: 17–30.

Derkzen, P., Franklin, A. and Bock, B. (2008) Examining power struggles as a signifier of successful partnership working: a case study of partnership dynamics, *Journal of Rural Studies*, 24: 458–66.

Desmarais, A. (2008) The power of peasants: reflections on the meanings of La Vía Campesia, *Journal of Rural Studies*, 24: 138–49.

D'Haese, M., Verbeke, W., van Huylenbroeck, G., Kirsten, J. and D'Haese, L. (2005) New institutional arrangements for rural development: the case of local woolgrowers' associations in the Transkei area, South Africa, *Journal of Development Studies*, 41: 1444–66.

Dibden, J. and Cocklin, C. (2005) Sustainability and agri-environmental governance, in Higgins, V. and Lawrence, G. (eds) *Agricultural Governance: Globalization and the New Politics of Regulation*. London: Routledge.

Dixon, D.P. and Hapke, H.M. (2003) Cultivating discourse: the social construction of agricultural legislation, *Annals of the Association of American Geographers*, 93: 142–64.

DuPuis, E.M. and Goodman, D. (2005) Should we go 'home' to eat? toward a reflexive politics of localism, *Journal of Rural Studies*, 21: 359–72.

Eakin, H. and Appendini, K. (2008) Livelihood change, farming, and managing flood risk in the Lerma Valley, Mexico, *Agriculture and Human Values*, 25: 555–66.

Echánove, F. (2005) Globalization and restructuring in rural Mexico: the case of fruit growers, *Tidjschrift voor Econimische en Sociale Geografie*, 96: 15–30.

Edensor, T. (2000) Walking in the British Countryside: Reflexivity, embodied practices and ways to escape, *Body and Society*, 6: 81–106.

Edensor, T. (2006) Performing rurality, in Cloke, P., Marsden, T. and Mooney, P. (eds) *Handbook of Rural Studies*. London: Sage.

Edwards, B., Goodwin M., Pemberton, S. and Woods, M. (2001) Partnerships, power and scale in rural governance, *Environment and Planning C: Government and Policy*, 19: 289–310.

Engel, S., Pagiola, S. and Wunder, S. (2008) Designing payments for environmental services in theory and practice: an overview of the issues, *Ecological Economics*, 65: 663–74.

Englund, H. (2002) The village in the city, the city in the village: migrants in Lilongwe, *Journal of Southern African Studies*, 28: 137–54.

Enticott, G. (2001) Calculating nature: the case of badgers, bovine tuberculosis and cattle, *Journal of Rural Studies*, 17: 149–64.

Epp, R. and Whitson, D. (eds) (2001) *Writing Off the Rural West*. Edmonton: University of Alberta Press.

Errington, A. (1997) Rural employment issues in the periurban fringe, in Bollman, R.D. and Bryden, J.D. (eds) *Rural Employment: an International Perspective*. Wallingford: CABI.

Evans, N., Morris, C. and Winter, M. (2002) Conceptualizing agriculture: a critique of post-productivism as the new orthodoxy, *Progress in Human Geography*, 26: 313–32.

Everett, S. and Aitchison, C. (2008) The role of food tourism in sustaining regional identity: a case study of Cornwall, South West England, *Journal of Sustainable Tourism*, 16: 150–67.

Eversole, R. and Martin, J. (2006) Jobs in the Bush: Global industries and inclusive rural development, *Social Policy and Administration*, 40: 692–704.

Falk, W.W., Hunt, L.L. and Hunt, M.O. (2004) Return migrations of African-Americans to the South: Reclaiming a land of promise, going home, or both? *Rural Sociology*, 69: 490–509.

Fan, C., Wall, G. and Mitchell, C.J.A. (2008) Creative destruction and the water town of Luzhi, China, *Tourism Management*, 29: 648–60.

Farinelli, B. (2008) *L'avenir est a la Campagne*. Paris: Syros.

Featherstone, D. (2003) Spatialities of transnational resistance to globalization: maps of grievance of the Inter-Continental Caravan, *Transactions of the Institute of British Geographers*, 28: 404–21.

Featherstone, D. (2008) *Resistance, Space and Political Identities: the making of counter-global networks*. London: Wiley-Blackwell.

Flognfeldt, T. (2006) Second homes, work commuting and amenity migrants in Norway's mountain areas, in Moss, L.A.G. (ed.) *The Amenity Migrants*. Wallingford: CABI.

Flora, C.B., Flora, J.L. and Fey, S. (2008) *Rural Communities: Legacy and Change*, 3rd edn, Boulder: Westview.

Fonte, M. (2008) Knowledge, food and place: a way of producing, a way of knowing, *Sociologia Ruralis*, 48: 200–22.

Foucault, M. (1976) *The Birth of the Clinic*. London: Tavistock.

Foucault, M. (1978) *The History of Sexuality*, vol. 1. London: Penguin.

Foucault, M. (1991) Governmentality, in Burchell, G., Gordon, C. and Miller, P. (eds) *The Foucault Effect: Studies in Governmentality*. Hemel Hempstead: Harvester Wheatsheaf.

Frank, E. (1999) *Gender, Agricultural Development and Food Security in Amhara, Ethiopia: The contested identity of women farmers in Ethiopia*. USAID Ethiopia. Available http: <http:\\pdf.dec.org/pdf_docs/Pnacg552.pdf>

Frankenberg, R. (1966) *Communities in Britain: Social life in town and country*. Harmondsworth: Penguin.

Frost, P.G.H. and Bond, I. (2008) The CAMPFIRE programme in Zimbabwe: payments for wildlife services, *Ecological Economics*, 65: 776–87.

Galeano, E. (2009) *Open Veins of Latin America*. London: Serpent's Tail. [First published 1973.]

Gallent, N., Juntti, M., Kidd, S. and Shaw, D. (2008) *Introduction to Rural Planning*. London: Routledge.

Gallent, N., Mace, A. and Tewdwr-Jones, M. (2005) *Second Homes: European perspectives and UK policies*. Aldershot: Ashgate.

Garland, J. and Chakraborti, N. (2004) Another country? Community, belonging and exclusion in rural England, in Chakraborti, R. and Garland, J. (eds) *Rural Racism*. Cullompton, UK: Willan.

Garrod, B., Wornell, R. and Youell, R. (2006) Re-conceptualising rural resources as countryside capital: the case of rural tourism, *Journal of Rural Studies*, 22: 117–28.

Gedicks, A. (2001) *Resource Rebels: Native challenges to mining and oil corporations*. Cambridge, MA: South End Press.

Gerrard, S. (2008) A travelling fishing village: the specific conjunctions of place, in Bærenholdt, J.O. and Granås, B. (eds) *Mobility and Place: Enacting Northern European Peripheries*. Aldershot: Ashgate.

Gezon, L.L. (2006) *Global Visions, Local Landscapes: A political ecology of conservation, conflict and control in northern Madagascar*. Lanham, MD: Alta Mira.

Gibson, C. and Davidson, D. (2004) Tamworth, Australia's 'country music capital': place marketing, rurality and resident reactions, *Journal of Rural Studies*, 20: 387–404.

Gibson, C.C. and Marks, S.A. (1995) Transforming rural hunters into conservationists: an assessment of community-based wildlife management programs in Africa, *World Development*, 23: 941–57.

Gombay, N. (2005) Shifting identities in a shifting world: food, place, community, and the politics of scale in an Inuit settlement, *Environment and Planning D: Society and Space*, 23: 415–33.

González, G.G. (1994) *Labor and Community: Mexican Citrus Worker Villages in a Southern California County, 1900–1950*. Chicago: University of Illinois Press.

Goodman, D.E. and Redclift, M. (eds) (1989) *The International Farm Crisis*, London: Macmillan.

Goodman, D.E. and Redclift, M. (1991) *Refashioning Nature: Food, ecology and culture*, London: Routledge.

Goodman, D.E., Sorj, B. and Wilkinson, J. (1987) *From Farming to Biotechnology: A theory of agro-industrial development*, Oxford: Blackwell.

Goodwin, M. (1998) The governance of rural areas: some emerging research issues and agendas, *Journal of Rural Studies*, 14: 5–12.

Gorman-Murray, A., Waitt, G. and Gibson, C. (2008) A queer country? A case study of the politics of gay/lesbian belonging in an Australian country town, *Australian Geographer*, 39: 171–91.

Gouveia, L. and Juska, A. (2002) Taming nature, taming workers: constructing the separation between meat consumption and meat production in the US, *Sociologia Ruralis*, 42: 370–90.

Gray, J. (2000) The Common Agricultural Policy and the re-invention of the rural in the European Community, *Sociologia Ruralis*, 40: 30–52.

Green, B. (1996) *Countryside Conservation*. London: E&FN Spon.

Gregory, D. (1994) Discourse, in Johnston, R.J., Gregory, D. and Smith, D.M. (eds) *The Dictionary of Human Geography*. Oxford: Blackwell.

Grinspun, R. (2003) Exploring the links among global trade, industrial agriculture and rural underdevelopment, in North, L.L. and Cameron, J.D. (eds) *Rural Progress, Rural Decay: neoliberal adjustment policies and local initiatives*. Bloomfield, CT: Kumarian Press.

Guthman, J. (2002) Commodified meanings, meaningful commodities: re-thinking production-consumption links through the organic system of provision, *Sociologia Ruralis*, 42: 295–311.

Guthman, J. (2004) The trouble with 'organic lite' in California: a rejoinder to the 'conventionalisation' debate, *Sociologia Ruralis*, 3: 301–17.

Gutman, P. (2007) Ecosystem services: foundations for a new rural-urban compact, *Ecological Economies*, 62: 383–87.

Halfacree, K. (1993) Locality and social representation: space, discourse and alternative definitions of the rural, *Journal of Rural Studies*, 9: 1–15.

Halfacree, K. (1994) The importance of 'the rural' in the constitution of counter-urbanization: evidence from England in the 1980s, *Sociologia Ruralis*, 34: 164–89.

Halfacree, K. (1999) A new space or spatial effacement? Alternative futures for the post-productivist countryside, in Walford, N., Everitt, J. and Napton, D. (eds) *Reshaping the Countryside: Perceptions and processes of rural change*. Wallingford: CAB International.

Halfacree, K. (2006) Rural space: constructing a three-fold architecture, in Cloke, P., Marsden, T. and Mooney, P. (eds) *Handbook of Rural Studies*. London: Sage.

Halfacree, K. (2007) Trial by space for a 'radical rural': introducing alternative localities, representations and lives, *Journal of Rural Studies*, 23: 125–41.

Halfacree, K. (2008) To revitalise counterurbanisation research? Recognising an international and fuller picture, *Population, Space and Place*, 14: 479–95.

Halhead, V. (2006) Rural movements in Europe: Scandinavia and the accession states, *Social Policy and Administration*, 40: 596–611.

Hall, C. (2008) Identifying farmer attitudes towards genetically-modified (GM) crops in Scotland: Are they pro- or anti-GM? *Geoforum*, 39: 204–12.

Hall, P., Thomas, R., Gracey, H. and Drewett, R. (1973) *The Containment of Urban England*. London: Allen and Unwin.

Halpin, D. (2004) Transitions between formations and organisations: an historical perspective on the political representation of Australian farmers, *Australian Journal of Politics and History*, 50: 469–90.

Halseth, G. and Rosenberg, M. (1995) Complexity in the rural Canadian housing landscape, *The Canadian Geographer*, 39: 336–52.

Halseth, G. and Sullivan, L. (2002) *Building Community in an Instant Town*. Prince George, BC: University of Northern British Columbia Press.

Hanbury-Tenison, R. (1997) 'Life in the Countryside', *Geographical Magazine*, November (sponsored feature).

Haraway, D. (1991) *Simians, Cyborgs and Women*. London: Free Association.

Harper, J. (2005) 'Press wilfully ignorant of US rural life', *The Washington Times*, 11 April.

Harper, S. (1988) Rural reference groups and images of place, in Pocock, D. (ed.) *Humanistic Approaches in Geography*. University of Durham, Department of Geography, Occasional Publication 22.

Harrington, L. (2005) Vulnerability and sustainability concerns for the U.S. high plains, in Essex, S., Gilg, A., Yarwood, R., Smithers, J. and Wilson, R. (eds) *Rural Change and Sustainability: Agriculture, Environment and Communities*. Wallingford: CABI.

Harrington, L. (2010) The U.S. Great Plains, Change, and Place Development, in Halseth, G., Markey, S. and Bruce, D. (eds) *The Next Rural Economies: Constructing rural place in global economies*. Wallingford: CABI.

Hart, G. (1991) Engendering everyday resistance: gender, patronage and production politics in rural Malaysia, *Journal of Peasant Studies*, 19: 93–121.

Hart, J.F. (1974) *The Look of the Land*. Engelwood Cliffs, CA: Prentice Hall.

Harvey, D. (1985) *The Urbanization of Capital: Studies in the History and Theory of Capitalist Urbanization*. Baltimore: Johns Hopkins University Press.

Harvey, D. (2005) *A Brief History of Neoliberalism*. Oxford: Oxford University Press.

Harvey, G. (1998) *The Killing of the Countryside*. London: Vintage.

Harvie, R. and Jobes, P.C. (2001) Social control of vice in post-frontier Montana, *Journal of Rural Studies*, 17: 235–46.

Hayami, Y. (2004) An ecological and historical perspective on agricultural development in Southeast Asia, in Akiyama, T. and Larson, D.F. (eds) *Rural Development and Agricultural Growth in Indonesia, the Philippines and Thailand*. Canberra: Asia Pacific Press.

Hayden, D. (2004) *A Field Guide to Sprawl*. New York: Norton.

Heley, J. (2008) Rounds, Range Rovers and rurality: the drinking geographies of a New Squirearchy, *Drugs: Education, Prevention and Policy*, 15: 315–21.

Heley, J. (2010) The new squirearchy and emergent cultures of the new middle classes in rural areas, *Journal of Rural Studies*, in press.

Henderson, G.L. (1998) *California and the Fictions of Capital*. Philadelphia: Temple University Press.

Hendrickson, M. and Heffernan, W.D. (2002) Opening spaces through relocalisation: locating potential resistance in the weaknesses of the global food system, *Sociologia Ruralis*, 42: 347–69.

Herbert-Cheshire, L. (2003) Translating policy: power and action in Australia's country towns. *Sociologia Ruralis*, 43: 454–73.

Herzog, T.R. and Barnes, G.J. (1999) Tranquility and preference revisted, *Journal of Environmental Psychology*, 19: 171–81.

Hettne, B. (1995) *Development Theory and the Three Worlds*. Harlow: Longman.

Higgins, V. and Lawrence, G. (eds) (2005) *Agricultural Governance: Globalization and the new politics of regulation*. London: Routledge.

High, C. and Nemes, G. (2007) Social learning in LEADER: Exogenous, endogenous and hybrid evaluation in rural development, *Sociologia Ruralis*, 47: 103–20.

Hinrichs, C.C. (2003) The practice and politics of food system localization, *Journal of Rural Studies*, 19: 33–46.

Hogan, J. (2004) Constructing the global in two rural communities in Australia and Japan, *Journal of Sociology*, 40: 21–40.

Hoggart, K. (1990) Let's do away with rural, *Journal of Rural Studies*, 6: 245–57.

Hoggart, K. (ed.) (2005) *The City's Hinterland: Dynamism and divergence in Europe's peri-urban territories*. Aldershot: Ashgate.

Hoggart, K. and Henderson, S. (2005) Excluding exceptions: housing non-affordability and the oppression of environmental sustainability, *Journal of Rural Studies*, 21: 181–96.

Holland, P. and Mooney, B. (2006) Wind and water: Environmental learning in early colonial New Zealand, *New Zealand Geographer*, 62: 39–49.

Holland, P., O'Connor, K. and Wearing, A. (2002) Remaking the grasslands of the open country, in Pawson, E. and Brooking, T. (eds) *Environmental Histories of New Zealand*. Melbourne: Oxford University Press.

Hollander, G.M. (2004) Agricultural trade liberalization, multifunctionality and sugar in the south Florida landscape, *Geoforum*, 35: 299–312.

Holloway, J. (2003) Spiritual embodiment and sacred rural landscapes, in Cloke, P. (ed.) *Country Visions*. Harlow: Pearson.

Holloway, L. (2004) Showing and telling farming: agricultural shows and re-imagining British agriculture, *Journal of Rural Studies*, 20: 319–30.

Holloway, L. and Kneafsey, M. (2000) Reading the space of the farmers' market: a case study from the United Kingdom, *Sociologia Ruralis*, 40: 285–99.

Holloway, L. and Kneafsey, M. (2004) Producing-consuming food: closeness, connectedness and rurality in four 'alternative' food networks, in Holloway, L. and Kneafsey, M. (eds) *Geographies of Rural Cultures and Societies*. Aldershot: Ashgate.

Holloway, S. (2004) Rural roots, rural routes: discourses of rural self and travelling other in debates about the future of Appleby New Fair, 1945–69, *Journal of Rural Studies*, 20: 143–56.

Holloway, S. (2005) Articulating otherness? White rural residents talk about Gypsy-Travellers, *Transactions of the Institute of British Geographers*, 30: 351–67.

Holmes, J. (2006) Impulses towards a multifunctional transition in rural Australia: gaps in the research agenda, *Journal of Rural Studies*, 22: 142–60.

Hong, K. (2001) The geography of time and labor in the late antebellum American rural south: *Fin-de*-servitude time consciousness, contested labor, and plantation capitalism, *International Review of Social History*, 46: 1–27.

Hubbard, P. (2005) 'Inappropriate and incongruous': opposition to asylum centres in the English countryside, *Journal of Rural Studies*, 21: 3–18.

Hubbard, P. (2006) *City*. London: Routledge.

Hughes, A. (1997) Rurality and 'cultures of womanhood', in Cloke, P. and Little, J. (eds) *Contested Countryside Cultures*. London: Routledge.

Hughes, G. (1992) Tourism and the geographical imagination, *Leisure Studies*, 11: 31–42.

IFOAM (2007) *The World of Organic Agriculture: Statistics and Emerging Trends 2007*. Bonn: International Federation of Organic Agriculture Movements.

Ilbery, B. (ed.) (1998) *The Geography of Rural Change*. Harlow: Addison Wesley Longman.

Ilbery, B. and Bowler, I. (1998) From agricultural productivism to post-productivism, in Ilbery, B. (ed.) *The Geography of Rural Change*. Harlow: Addison Wesley Longman.

Ilbery, B. and Maye, D. (2008) Placing local food in a cross-border setting, in Stringer, C. and Le Heron, R. (eds) *Agri-Food Commodity Chains and Globalising Networks*. Aldershot: Ashgate.

Inglis, T. (2008) *Global Ireland*. London and New York: Routledge.

Ingold, T. (1993) The temporality of landscape, *World Archaeology*, 25: 152–74.

Ingold, T. (1995) Building, dwelling, living: how people and animals make themselves at home in the world, in Strathern, M. (ed.) *Shifting Contexts: Transformations in Anthropological Knowledge*. London: Routledge.

ISAAA (2007) *ISAAA Briefing 37 – 2007*. Ithaca, NY: Information Service on the Acquisition of Agricultural Biotechnology Applications. Available http: <www.isaaa.org>

Isserman, A.M., Fraser, E. and Warren, D.E. (2009) Why some rural places prosper and others do not, *International Regional Science Review*, 32: 300–42.

Jackiewicz, E.L. (2006) Community-centred globalization: modernization under control in rural Costa Rica, *Latin American Perspectives*, 33: 136–46.

Jarosz, L. (2008) The city in the country: growing alternative food networks in metropolitan areas, *Journal of Rural Studies*, 24: 231–44.

Jauhiainen, J.S. (2009) Will the retiring baby boomers return to rural periphery? *Journal of Rural Studies*, 25: 23–34.

Jewitt, S. and Baker, K. (2007) The Green Revolution re-assessed: Insider perspectives on agrarian change in Bulandshahr District, Western Uttar Pradesh, India, *Geoforum*, 38: 73–89.

Johnsen, S. (2003) Contingency revealed: New Zealand farmers' experiences of agricultural restructuring, *Sociologia Ruralis*, 43: 128–53.

Jones, K.R. and Wills, J. (2005) *The Invention of the Park*. Malden, MA: Polity.

Jones, O. (1995) Lay discourses of the rural: developments and implications for rural studies, *Journal of Rural Studies*, 11: 35–49.

Jones, O. (1997) Little figures, big shadows: country childhood stories, in Cloke, P. and Little, J. (eds) *Contested Countryside Cultures*. London: Routledge.

Jones, O. (2003) 'The restraint of beasts': rurality, animality, Actor Network Theory and dwelling, in Cloke, P. (ed.) *Country Visions*. London: Arnold.

Jordan, J.A. (2007) The heirloom tomato as cultural object: investigating taste and space, *Sociologia Ruralis*, 47: 20–41.

Jowitt, J. (2008) Shoppers lose their taste for organic food, *The Guardian*, 29 August, p. 1.

Juska, A. (2007) Discourses on rurality in post-socialist news media: The case of Lithuania's leading daily 'Lietuvos Rytas' (1991–2004), *Journal of Rural Studies*, 23: 238–53.

Kadigi, R., Mdoe, N. and Ashimogo, G. (2007) Understanding poverty through the eyes of the poor: the case of Usanga Plains in Tanzania, *Physics and Chemistry of the Earth*, 32: 1330–38.

Kapoor, I. (2002) The Devil's in the theory: a critical assessment of Robert Chambers' work on participatory development, *Third World Quarterly*, 23: 101–17.

Kawagoe, T. (2004) The political economy of rural development in Indonesia, in Akiyama, T. and Larson, D.F. (eds) *Rural Development and Agricultural Growth in Indonesia, the Philippines and Thailand*. Canberra: Asia Pacific Press.

Kimura, A.H. and Nishiyama, M. (2008) The *chisan-chiso* movement: Japanese local food movement and its challenges, *Agriculture and Human Values*, 25: 49–64.

Kitchen, L. and Marsden, T. (2009) Creating sustainable rural development through stimulating the eco-economy: beyond the eco-economic paradox? *Sociologia Ruralis*, 49: 273–94.

Kneafsey, M., Ilbery, B. and Jenkins, T. (2001) Exploring the dimensions of culture economies in rural West Wales, *Sociologia Ruralis*, 41: 296–310.

Kneen, B. (2002) *Invisible Giant: Cargill and its transnational strategies*. London: Pluto Press.

Knobloch, F. (1996) *The Culture of Wilderness*. Chapel Hill: University of North Carolina Press.

Koczberski, G. (2007) Loose fruit mamas: creating incentives for smallholder women in oil palm production in Papua New Guinea, *World Development*, 35: 1172–85.

Kolodny, A. (1975) *The Lay of the Land: Metaphor as experience in American Life and Letters*. Chapel Hill: University of North Carolina Press.

Kontuly, T. (1998) Contrasting the counterurbanisation experience in European nations, in Boyle, P. and Halfacree, K. (eds) *Migration to Rural Areas*. Chichester: Wiley.

Korf, B. and Oughton, E. (2006) Rethinking the European countryside – can we learn from the South? *Journal of Rural Studies*, 22: 278–89.

Kovacs, I. and Kucerova, E. (2006) The project class in central Europe, *Sociologia Ruralis*, 46: 3–21.

Kurtz, M. and Craig, V. (2009) Constructing rural geographies in publication, *ACME: An International E-Journal for Critical Geographies*, 8: 376–93.

Lacour, C. and Puissant, S. (2007) Re-urbanity: urbanising the rural and ruralising the urban, *Environment and Planning A*, 39: 728–47.

Lapping, M.B., Daniels, T.L. and Keller, J.W. (1989) *Rural Planning and Development in the United States*. New York: Guilford Press.

Larner, W. (1998) Hitching a ride on the tiger's back: globalisation and spatial imaginaries in New Zealand, *Environment and Planning D: Society and Space*, 16: 599–614.

Larsen, S., Sorenson, C., McDermott, D., Long, J. and Post, C. (2007) Place perception and social interaction on an exurban landscape in central Colorado, *Professional Geographer*, 59: 421–33.

Latour, B. (1993) *We Have Never Been Modern*. Hemel Hempstead: Harvester Wheatsheaf.

Lawson, V., Jarosz, L. and Bonds, A. (2008) Building economies from the bottom up: (mis)representations of poverty in the rural American Northwest, *Social and Cultural Geography*, 9: 737–53.

Leeuwis, C. (2000) Reconceptualising participation for sustainable rural development: towards a negotiation approach, *Development and Change*, 31: 939–59.

Lefebvre, H. (1991) *The Production of Space*. Oxford: Blackwell.

Le Heron, R. (1993) *Globalised Agriculture*. London: Pergamon.

Leyshon, M. (2008) The betweeness of being a rural youth: inclusive and exclusive lifestyles, *Social and Cultural Geography*, 9: 1–26.

Liang, Z., Chen, Y.P. and Gu, Y. (2002) Rural industrialisation and internal migration in China, *Urban Studies*, 39: 2175–87.

Lichter, D.T. and Johnson, K.M. (2007) The changing spatial concentration of America's rural poor population, *Rural Sociology*, 72: 331–58.

Liepins, R. (2000a) New energies for an old idea: reworking approaches to 'community' in contemporary rural studies, *Journal of Rural Studies*, 16: 23–36.

Liepins, R. (2000b) Exploring rurality through 'community': discourses, practices and spaces shaping Australian and New Zealand rural 'communities', *Journal of Rural Studies*, 16: 325–42.

Liepins, R. (2000c) Making men: the construction and representation of agriculture-based masculinities in Australia and New Zealand, *Rural Sociology*, 65: 605–20.

Liepins, R. and Bradshaw, B. (1999) Neo-liberal agricultural discourse in New Zealand: Economy, culture and politics linked, *Sociologia Ruralis*, 39: 563–82.

Little, J. (2002) *Gender and Rural Geography*. Harlow: Pearson.

Little, J. (2003) Riding the rural love train: heterosexuality and the rural community, *Sociologia Ruralis*, 43: 401–17.

Little J., Ilbery, B. and Watts, D. (2009) Gender, consumption and the relocalisation of food: a research agenda, *Sociologia Ruralis*, 49: 201–17.

Little, J.I. (2009) Scenic tourism on the northeastern borderland: Lake Memphremagog's steamboat excursions and resort hotels, 1850–1900, *Journal of Historical Geography*, 35: 716–42.

Livingston, J.A. (1996) Other selves, in Vitek, W. and Jackson, W. (eds) *Rooted in the Land*. New Haven: Yale University Press.

Lobely, M., Potter, C., Butler, A., Whitehead, I. and Millard, N. (2005) *The Wider Social Impact of Changes in the Structure of Agricultural Businesses*. Exeter: Centre for Rural Research.

Long, H. (2007) Brief introduction of China's policy on 'Building a New Countryside', *Proceedings of China-France International Symposium on Rural Construction and Development*, 87–88.

Long, H., Liu, Y., Wu, X. and Dong, G. (2009) Spatial-temporal dynamic patterns of farmland and rural settlements in Su-Xi-Chang region: implications for building a new countryside in coastal China, *Land Use Policy*, 26: 322–33.

Lowe, P., Clark, J., Seymour, S. and Ward, N. (1997) *Moralizing the Environment: Countryside change, farming and pollution*. London: UCL Press.

Lowe, P. and Goyder, G. (1983) *Environmental Groups in Politics*. London: Allen and Unwin.

Lowe, P. and Phillipson, J. (2006) Reflexive inter-disciplinary research: the making of a research programme on rural economy and land use, *Journal of Agricultural Economics*, 57: 165–84.

Lynch, K. (2005) *Rural-Urban Interactions in the Developing World*. London: Routledge.

MacDonald, S.A. (2005) *The Agony of an American Wilderness: Loggers, environmentalists and the struggle for control of a forgotten forest*. Lanham, MD: Rowman and Littlefield.

Mackenzie, A.F.D. (2004) Re-imagining the land, North Sutherland, Scotland, *Journal of Rural Studies*, 20: 273–87.

Mackenzie, A.F.D. (2006a) "S Leinn Fhèin am Fearann" (The land is ours): reclaiming land, re-creating community, North Harris, Outer Hebrides, Scotland, *Environment and Planning D: Society and Space*, 24: 577–98.

Mackenzie, A.F.D. (2006b) A working land: crofting communities, place and the politics of the possible in post-Land Reform Scotland, *Transactions of the Institute of British Geographers*, 31: 383–98.

Macnaghten, P. and Urry, J. (1998) *Contested Natures*. London and Thousand Oaks, CA: Sage.

MacPherson, H. (2009) The intercorporeal emergence of landscape: negotiating sight, blindness, and ideas of landscape in the British countryside, *Environment and Planning A*, 41: 1042–54.

Magnani, N. and Struffi, L. (2009) Translation sociology and social capital in rural development initiatives: a case study from the Italian Alps, *Journal of Rural Studies*, 25: 231–38.

Magnusson, W. and Shaw, K. (eds) (2003) *A Political Space: Reading the global through Clayoquot Sound*. Minneapolis: University of Minnesota Press.

Markey, S. (2010) Fly-in, fly-out resource development: a new regionalist perspective on the next rural economy, in Halseth, G., Markey, S. and Bruce, D. (eds) *The Next Rural Economies: Constructing rural place in global economies*. Wallingford: CABI.

Markey, S., Halseth, G. and Manson, D. (2008) Challenging the inevitability of rural decline: advancing the policy of place in northern British Columbia, *Journal of Rural Studies*, 24: 409–21.

Marsden, T. (1999) Rural futures: the consumption countryside and its regulation, *Sociologia Ruralis*, 39: 501–20.

Marsden, T. (2003) *The Condition of Rural Sustainability*. Assen, NL: Van Gorcum.

Marston, S., Jones, J.P. and Woodward, K. (2005) Human geography without scale, *Transactions of the Institute of British Geographers*, 30: 416–32.

Marvin, G. (2000) The problem of foxes: legitimate and illegitimate killing in the English countryside, in Knight, J. (ed.) *Natural Enemies: People-wildlife conflicts in anthropological perspective*. London: Routledge.

Marvin, G. (2003) A passionate pursuit: foxhunting as performance, in Szerszynski, B., Heim, W. and Waterton, C. (eds) *Nature Performed: Environment, culture and performance*. Oxford: Blackwell.

Massey, D. (2004) Geographies of responsibility, *Geografiska Annaler B*, 86: 5–18.

Massey, D. (2005) *For Space*. London: Sage.

Masuda, J.R. and Garvin, T. (2008) Whose Heartland? The politics of place in a rural-urban interface, *Journal of Rural Studies*, 24: 112–23.

Mather, A.S., Hill, G. and Nijnik, M. (2006) Post-productivism and rural land use: cul de sac or challenge for theorization, *Journal of Rural Studies*, 22: 441–55.

Matless, D. (1994) Doing the English village, 1945–90: an essay in imaginative geography, in Cloke, P., Doel, M., Matless, D., Phillips, M. and Thrift, N., *Writing The Rural*. London: Paul Chapman.

Matless, D. (1998) *Landscape and Englishness*. London: Reaktion Books.

Matless, D. (2005) Sonic geography in a nature region, *Social and Cultural Geography*, 6: 745–66.

Matsinos, Y.G., Mazaris, A.D., Papadimitrou, K.D., Mniestris, A., Hatzigiannidis, G., Maioglou, D. and Pantis, J.D. (2008) Spatio-temporal variability in human and natural sounds in a rural landscape, *Landscape Ecology*, 23: 945–59.

Maye, D., Ilbery, B. and Kneafsey, M. (2005) Changing places: investigating the cultural terrain of village pubs in south Northamptonshire, *Social and Cultural Geography*, 6: 831–47.

McAfee, K. (2008) Beyond techno-science: transgenic maize in the fight over Mexico's future, *Geoforum*, 39: 148–60.

McCalla, A.F. and Nash, J. (eds) (2007) *Reforming Agricultural Trade for Developing Countries. Vol. 1: Key Issues for a Pro-Development Outcome of the Doha Round*. Washington DC: IBRD/The World Bank.

McCarthy, J. (2004) Privatizing conditions of production: trade agreements as neoliberal environmental governance, *Geoforum*, 35: 327–41.

McDonald, J.H. (2001) Reconfiguring the countryside: power, control, and the (re)organization of farmers in west Mexico, *Human Organization*, 60: 247–58.

McGregor, J. (2005) Crocodile crimes: people versus wildlife and the politics of postcolonial conservation on Lake Kariba, Zimbabwe, *Geoforum*, 36: 353–69.

Meijering, L., Huigen, P. and Van Hoven, B. (2007) Intentional communities in rural spaces, *Tijdschrift voor Economische en Sociale Geografie*, 98: 42–52.

Merrifield, A. (2000) Henri Lefebvre: a socialist in space, in Crang, M. and Thrift, N. (eds) *Thinking Space*, London: Routledge.

Merriman, P. (2005) 'Respect the life of the countryside': the Country Code, government and the conduct of visitors to the countryside in post-war England and Wales, *Transactions of the Institute of British Geographers*, 30: 336–50.

Mertz, O., Mbow, C., Reenberg, A. and Diuof, A. (2009) Farmers' perceptions of climate change and agricultural adaptation strategies in rural Sahel, *Environmental Management*, 43: 804–16.

Miele, M. and Murdoch, J. (2002) The practical aesthetics of traditional cuisines: slow food in Tuscany, *Sociologia Ruralis*, 42: 312–28.

Milbourne, P. (2003) The complexities of hunting in England and Wales, *Sociologia Ruralis*, 43: 289–308

Milbourne, P. (2004) *Rural Poverty: Marginalisation and exclusion in Britain and the United States*. London: Routledge.

Milbourne, P., Mitra, B. and Winter, M. (2001) *Agriculture and Rural Society*. London: Defra.

Millard, A.V. and Chapa, J. (2004) *Apple Pie and Enchiladas: Latino newcomers in the rural Midwest*. Austin: University of Texas Press.

Millstone, E. and Lang, T. (2003) *The Atlas of Food*. London: Earthscan.

Mitchell, C.J.A. (1998) Entrepreneurialism, commodification and creative destruction: a model of post-modern community development, *Journal of Rural Studies*, 14: 273–86.

Mitchell, C.J.A. (2004) Making sense of counterurbanization, *Journal of Rural Studies*, 20: 15–34.

Mitchell, C.J.A. and de Waal, S.B. (2009) Revisiting the model of creative destruction: St Jacobs, Ontario, a decade later, *Journal of Rural Studies*, 25: 156–67.

Mitchell, D. (1996) *The Lie of the Land: Migrant workers and the California landscape*. Minneapolis: University of Minnesota Press.

Mitchell, D. (2000) Dead labour and the political economy of landscape – California Living, California Dying, in Anderson, K., Domosh, M. and Pile, S. (eds) *Handbook of Cultural Geography*, London: Sage.

Monbiot, G. (1994) *No Man's Land: An investigative journey through Kenya and Tanzania*. London: Picador.

Mordue, T. (1999) Heartbeat country: conflicting values, coinciding visions, *Environment and Planning A*, 31: 629–46.

Morgan, K. and Murdoch, J. (2000) Organic vs conventional agriculture: knowledge, power and innovation in the food chain, *Geoforum*, 31: 159–73.

Mormont, M. (1987) The emergence of rural struggles and their ideological effects, *International Journal of Urban and Regional Research*, 7: 559–78.

Mormont, M. (1990) Who is rural? Or, how to be rural: Towards a sociology of the rural, in Marsden, T., Lowe, P. and Whatmore, S. (eds), *Rural Restructuring: Global Processes and their Responses*. London: David Fulton.

Morris, C. and Evans, N. (2004) Agricultural turns, geographical turns: retrospect and prospect, *Journal of Rural Studies*, 20: 95–111.

Moschini, G. (2008) Biotechnology and the development of food markets: retrospect and prospects, *European Review of Agricultural Economics*, 35: 331–55.

Moseley, M. (2003) *Rural Development: Principles and Practice*. London: Sage.

Moss, L.A.G. (ed.) (2006) *The Amenity Migrants: Seeking and sustaining mountains and their cultures*. Wallingford: CABI.

Moyo, S. and Yeros, P. (eds) (2005) *Reclaiming the Land: The resurgence of rural movements in Africa, Asia and Latin America*. London and New York: Zed Books.

Mukherjee, A. and Zhang, X. (2007) Rural industrialization in China and India: role of policies and institutions, *World Development*, 35: 1621–34.

Murdoch, J. (1997a) Towards a geography of heterogeneous associations, *Progress in Human Geography*, 21: 321–37.

Murdoch, J. (1997b) The shifting territory of government: some insights from the Rural White Paper, *Area*, 29: 109–18.

Murdoch, J. (1998) The spaces of actor-network theory, *Geoforum*, 29: 357–74.

Murdoch, J. (2003) Co-constructing the countryside: hybrid networks and the extensive self, in Cloke, P. (ed.) *Country Visions*, Harlow: Pearson.

Murdoch, J. (2006) Networking rurality: emergent complexity in the countryside, in Cloke, P., Marsden, T. and Mooney, P. (eds) *Handbook of Rural Studies*, London: Sage.

Murdoch, J. and Abram, S. (2002) *Rationalities of Planning*. Aldershot: Ashgate.

Murdoch, J. and Lowe, P. (2003) The preservationist paradox: modernism, environmentalism and the politics of spatial division, *Transactions of the Institute of British Geographers*, 28: 318–32.

Murdoch, J. and Marsden, T. (1994) *Reconstituting Rurality*. London: UCL Press.

Murdoch, J. and Ward, N. (1997) Governmentality and territoriality: the statistical manufacture of Britain's 'national farm', *Political Geography*, 16: 307–24.

Murray, W. (2006a) Neo-feudalism in Latin America? Globalisation, agribusiness, and land re-concentration in Chile, *Journal of Peasant Studies*, 33: 646–77.

Murray, W. (2006b) *Geographies of Globalization*. London: Routledge.

Murton, J. (2007) *Creating a Modern Countryside*. Vancouver: University of British Columbia Press.

Narlikar, A. (2005) *The World Trade Organization: A Very Short Introduction*. Oxford: Oxford University Press.

Nash, L. (2004) The fruits of ill-health: pesticides and working bodies in post-World War II California, *Osiris*, 19: 203–19.

Neal, S. and Walters, S. (2007) 'You can get away with loads because there's no one here': Discourses of regulation and non-regulation in English rural spaces, *Geoforum*, 38: 252–63.

Neal, S. and Walters, S. (2008) Rural be/longing and rural social organizations: conviviality and community-making in the English countryside, *Sociology*, 42: 279–97.

Neefjes, K. (2000) *Environments and Livelihoods: Strategies for Sustainability*. Oxford: Oxfam

Neilson, J. and Pritchard, B. (2009) *Value Chain Struggles: Institutions and governance in the plantation districts of southern India*. London: Wiley – Blackwell.

Nelson, L. and Hiemstra, N. (2008) Latino immigrants and the renegotiation of place and belonging in small town America, *Social and Cultural Geography*, 9: 319–42.

Nelson, P.B. (2006) Geographic perspective on amenity migration across the USA: National-, regional- and local-scale analysis, in Moss, L.A.G. (ed.) *The Amenity Migrants*. Wallingford: CABI.

Nerlich, B. and Döring, M. (2005) Poetic justice? Rural policy clashes with rural poetry in the 2001 outbreak of foot and mouth disease in the UK, *Journal of Rural Studies*, 21: 165–80.

Newby, H. (1977) *The Deferential Worker: A Study of Farm Workers in East Anglia*. London: Allen Lane.

Newby, H., Bell, C., Rose, D. and Saunders, P. (1978) *Property, Paternalism and Power*. London: Hutchinson.

Ni Laoire, C. (2001) A matter of life and death? Men, masculinities and staying 'behind' in rural Ireland, *Sociologia Ruralis*, 41: 220–36.

Ni Laoire, C. (2007) The 'green grass of home'? Return migration to rural Ireland, *Journal of Rural Studies*, 23: 332–44.

North, D. (1998) Rural industrialization, in Ilbery, B. (ed.) *The Geography of Rural Change*. Harlow: Addison Wesley Longman.

Norton, A. (1996) Experiencing nature: the reproduction of environmental discourse through safari tourism in East Africa, *Geoforum*, 27: 355–73.

Opie, J. (1994) *The Law of the Land: Two hundred years of American farmland policy*. Lincoln: University of Nebraska Press.

Osbahr, H., Twyman, C., Adger, W.N. and Thomas, D.S.G. (2008) Effective livelihood adaptation to climate change disturbance: Scale dimensions of practice in Mozambique, *Geoforum*, 39: 1951–64.

Pallot, J. (1988) The USSR, in Cloke, P. (ed.) *Policies and Plans for Rural People: An International Perspective*. London: Unwin Hyman.

Panelli, R. (2006) Rural society, in Cloke, P., Marsden, T. and Mooney, P. (eds) *Handbook of Rural Studies*. London: Sage.

Panelli, R., Allen, D., Ellison, B., Kelly, A., John, A. and Tipa, G. (2008) Beyond Bluff oysters? Place identity and ethnicity in a peripheral coastal setting, *Journal of Rural Studies*, 24: 41–55.

Panelli, R., Nairn, K. and McCormack, J. (2002) 'We make our own fun': Reading the politics of youth with(in) community, *Sociologia Ruralis*, 42: 106–30.

Parker, G. (2006) The Country Code and the ordering of countryside citizenship, *Journal of Rural Studies*, 22: 1–16.

Paul, H. and Steinbrecher, R. (2003) *Hungry Corporations: Transnational biotech companies colonise the food chain*. London: Zed Books.

Pechlaner, G. and Otero, G. (2008) The third food regime: Neoliberal globalism and agricultural biotechnology in North America, *Sociologia Ruralis*, 48: 351–71.

Peet, R. (2003) *Unholy Trinity: The IMF, World Bank and WTO*. London and New York: Zed Books.

Peine, E. and McMichael, P. (2005) Globalization and global governance, in Higgins, V. and Lawrence, G. (eds) *Agricultural Governance: Globalization and the new politics of regulation*. London: Routledge.

Perkins, H.C. (2006) Commodification: re-resourcing rural areas, in Cloke, P., Marsden, T. and Mooney, P. (eds) *Handbook of Rural Studies*. London: Sage.

Perreault, T. (2001) Developing identities: indigenous mobilization, rural livelihoods and resource access in Ecuadorian Amazonia, *Ecumene*, 8: 381–413.

Perreault, T. (2005) State restructuring and the scale politics of rural water governance in Bolivia, *Environment and Planning A*, 37: 263–84.

Petras, J. (1997) Latin America: the resurgence of the Left, *New Left Review*, 223: 17–47.

Phillips, M. (2002) The production, symbolization and socialization of gentrification: impressions from two Berkshire villages, *Transactions of the Institute of British Geographers*, 27: 282–308.

Phillips, M. (2004) Obese and pornographic ruralities: further cultural twists for Rural Geography?, in Holloway, L. and Kneafsey, M. (eds) *Geographies of Rural Cultures and Societies*. Aldershot: Ashgate.

Phillips, M. (2008) Rurality as a globalised mediascope? Impressions from television drama production and distribution at the turn of the Millennium in Australia, Britain and New Zealand, *Critical Studies in Television*, 3: 16–44.

Phillips, M., Fish, R. and Agg, J. (2001) Putting together ruralities: towards a symbolic analysis of rurality in the British mass media, *Journal of Rural Studies*, 17: 1–28.

Phillips, S.T. (2007) *This Land, This Nation: Conservation, Rural America and the New Deal*. New York: Cambridge University Press.

Philo, C. (1992) Neglected rural geographies: a review, *Journal of Rural Studies*, 8: 193–207.

Philo, C. (1995) Animals, geography and the city: notes on inclusions and exclusions, *Environment and Planning D: Society and Space*, 13: 651–88.

Potter, C. (2004) Multifunctionality as an agricultural and rural policy concept, in Brouwer, F. (ed.) *Sustaining Agriculture and the Rural Environment*, Cheltenham: Edward Elgar.

Potter, C. and Burney, J. (2002) Agricultural multifunctionality in the WTO: legitimate non-trade concern or disguised protectionism? *Journal of Rural Studies*, 18: 35–47.

Potter, C. and Tilzey, M. (2005) Agricultural policy discourses in the European post-Fordist transition: neoliberalism, neomercantilism and multifunctionality, *Progress in Human Geography*, 29: 1–20.

Potter, R., Binns, T., Elliott, J.A. and Smith, D. (2008) *Geographies of Development*. Harlow: Pearson.

Pretty, G., Bramston, P., Patrick, J. and Pannach, W. (2006) The relevance of community sentiments to Australian rural youth's intention to stay in their home communities, *American Behavioral Scientist*, 50: 226–40.

Price, L. and Evans, N. (2009) From stress to distress: conceptualizing the British family farming patriarchal way of life, *Journal of Rural Studies*, 25: 1–11.

Prideaux, B. (2002) Creating rural heritage visitor attractions: the Queensland Heritage Trails Project, *International Journal of Tourism Research*, 4: 313–23.

Prudham, S. (2005) *Knock on Wood: Nature as commodity in Douglas-fir country*. New York: Routledge.

Prudham, S. (2008) Tall among the trees: Organizing against globalist forestry in rural British Columbia, *Journal of Rural Studies*, 24: 182–96.

Putnam, R.D. (1993) *Making Democracy Work: Civic Tradition in Modern Italy*. Princeton, NJ: Princeton University Press.

Putnam, R.D. (2000) *Bowling Alone: The Collapse and Renewal of American Community*. New York: Simon & Schuster.

Pyne, S.J. (2009) The human geography of fire: a research agenda, *Progress in Human Geography*, 33: 443–46.

Quarlman, D. (2001) Corporate hog farming: the view from the family farm, in Epp, R. and Whitson, D. (eds) *Writing Off the Rural West*. Edmonton: University of Alberta Press.

Ramírez-Ferrero, E. (2005) *Troubled Fields: Men, emotions and the crisis in American farming*. New York: Columbia University Press.

Ramp, W. and Koc, M. (2001) Global investment and local politics: the case of Lethbridge, in Epp, R. and Whitson, D. (eds) *Writing Off the Rural West*. Edmonton: University of Alberta Press.

Ray, C. (1998) Culture, intellectual property and territorial rural development, *Sociologia Ruralis*, 38: 3–19.

Ray, C. (1999) Endogenous development in the era of reflexive modernity, *Journal of Rural Studies*, 15: 257–67.

Ray, C. (2001) Transnational cooperation between rural areas: elements of a political economy of EU rural development, *Sociologia Ruralis*, 41: 279–95.

Ray, C. (2006) Neo-endogenous rural development in the EU, in Cloke, P., Marsden, T. and Mooney, P. (eds) *The Handbook of Rural Studies*, London: Sage.

Razavi, S. (2007) Liberalisation and the debates on women's access to land, *Third World Quarterly*, 28: 1479–1500.

Reed, M. (2008) The rural arena: the diversity of protest in rural England, *Journal of Rural Studies*, 24: 209–18.

Richards, P. (2004) Private versus public? Agenda-setting in international agro-technologies, in Jansen, K. and Vellema, S. (eds) *Agribusiness and Society*, London: Zed Books.

Richardson, T. (2000) Discourses of rurality in EU Spatial Policy: the European Spatial Development Perspective, *Sociologia Ruralis*, 40: 53–71.

Ricketts Hein, J., Ilbery, B. and Kneafsey, M. (2006) Distribution of local food activity in England and Wales: an index of food relocalization, *Regional Studies*, 40: 289–301.

Rigg, J. and Ritchie, M. (2002) Production, consumption and imagination in rural Thailand, *Journal of Rural Studies*, 18: 359–72.

Riley, M. (2009) Bringing the 'invisible farmer' into sharper focus: gender relations and agricultural practices in the Peak District (UK), *Gender, Place and Culture*, 16: 665–82.

Riley, M. and Harvey, D. (2007) Oral histories, farm practice and uncovering meaning in the countryside, *Social and Cultural Geography*, 8: 391–415.

Roberts, J. (1998) English gardens in India, *Garden History*, 26: 115–35.

Robertson, M. (2004) The neoliberalization of ecosystem services: wetland mitigation banking and problems in environmental governance, *Geoforum*, 35: 361–75.

Robinson, G. (2004) *Geographies of Agriculture*. Harlow: Pearson Prentice Hall.

Rogaly, B. (2006) Intensification of work-place regimes in British agriculture: the role of migrant workers, *Sussex Migration Working Papers 36*, Brighton: University of Sussex.

Rojek, C. and Urry, J. (1997) Transformations of travel and theory, in Rojek, C. and Urry, J. (eds) *Touring Cultures: Transformations of Travel and Theory*. London: Routledge.

Rose, G. (1993) *Feminism and Geography*. Cambridge: Polity Press.

Rose, N. (1996) The death of the social? Refiguring the territory of government, *Economy and Society*, 25: 327–56.

Routledge, P. (2003) Convergence space: process geographies of grassroots mobilization networks, *Transactions of the Institute of British Geographers*, 28: 333–49.

Routledge, P. and Cumbers, A. (2009) *Global Justice Networks: Geographies of transnational solidarity*. Manchester: Manchester University Press.

Rudy, A.P. (2005) Imperial contradictions: is the Valley a watershed, region or cyborg? *Journal of Rural Studies*, 21: 19–39.

Runte, A. (1997) *National Parks: The American Experience*. Lincoln, NE: University of Nebraska Press.

Rye, J.F. (2006) Rural youth's images of the rural, *Journal of Rural Studies*, 22: 409–21.

Rye, J.F. and Andrzejewska, J. (2010) The structural disempowerment of Eastern European migrant farm workers in Norwegian agriculture, *Journal of Rural Studies*, 26: 41–51.

Salamon, S. (2003) *Newcomers to Old Towns: Suburbanization of the Heartland*. Chicago: University of Chicago Press.

Saugeres, L. (2002a) Tractors and men: masculinity, technology and power, *Sociologia Ruralis*, 42: 143–59.

Saugeres, L. (2002b) Cultural representation of the farming landscape: masculinity, power and nature, *Journal of Rural Studies*, 18: 373–85.

Saunders, C., Kaye-Blake, W., Marshall, L., Greenhalgh, S. and Pereira, M.D. (2009) The impacts of a United States biofuel policy on New Zealand's agricultural sector, *Energy Policy*, 37: 3448–54.

Saville, J. (1957) *Rural Depopulation in England and Wales, 1851–1951*. London: Routledge and Kegan Paul.

Schivelbusch, W. (1986) *The Railway Journey: the industrialization of time and space in the 19th Century*. Berkeley: University of California Press.

Schmied, D. (2005) Incomers and locals in the European countryside, in Schmied, D. (ed.) *Winning and Losing: The changing geographies of Europe's rural areas*. Aldershot: Ashgate.

Schulman, M.D. and Anderson, C.D. (1999) The Dark Side of the Force: a case study of restructuring and social capital, *Rural Sociology*, 64: 351–72.

Scoones, I. (1998) Sustainable rural livelihoods: a framework of analysis, *IDS Working Paper 72*, Brighton: Institute of Development Studies.

Scoones, I. (2008) Mobilizing against GM crops in India, South Africa and Brazil, *Journal of Agrarian Change*, 8: 315–44.

Seijo, F. (2005) The politics of fire: Spanish forest policy and ritual resistance in Galicia, Spain, *Environmental Politics*, 14: 380–402.

Shambaugh-Miller, M. (2007) Development of a rural typology GIS for policy makers, paper presented to the Quadrennial Conference of British, Canadian and American Rural Geographers, Spokane, July.

Shaw Taylor, L. (2005) Family farms and capitalist farms in mid-nineteenth century England, *Agricultural History Review*, 53: 158–91.

Sheingate, A.D. (2001) *The Rise of the Agricultural Welfare State*. Princeton, NJ: Princeton University Press.

Shepherd, A. (1998) *Sustainable Rural Development*. Basingstoke: Macmillan.

Sherman, J. (2006) Coping with rural poverty: economic survival and moral capital in rural America, *Social Forces*, 85: 891–913.

Shigetomi, S. (2004) Rural organisations and development: the social background for collective action, in Akiyama, T. and Larson, D.F. (eds) *Rural Development and Agricultural Growth in Indonesia, the Philippines and Thailand*. Canberra: Asia Pacific Press.

Shiva, V. (1991) *The Violence of the Green Revolution: Third World agriculture, ecology and politics*. London: Zed Books.

Short, B. (2000) Rural demography, 1850–1914, in Collins, E.J.T. (ed.) *The Agrarian History of England and Wales VII: 1850–1914*. Cambridge: Cambridge University Press.

Short, B. (2006) Idyllic ruralities, in Cloke, P., Marsden, T. and Mooney, P. (eds) *Handbook of Rural Studies*, London: Sage.

Short, J.R. (1991) *Imagined Country*. London: Routledge.

Shortall, S. (2008) Are rural development programmes socially inclusive? Social inclusion, civic engagement, participation, and social capital: Exploring the differences, *Journal of Rural Studies*, 24: 450–57.

Shubin, S. (2007) Networked poverty in rural Russia, *Europe-Asia Studies*, 59: 591–620.

Shucksmith, M. (2000) Endogenous development, social capital and social inclusion: perspectives from LEADER in the UK, *Sociologia Ruralis*, 40: 208–18.

Sibley, D. (1997) Endangering the sacred: Nomads, youth cultures and the English countryside, in Cloke, P. and Little, J. (eds) *Contested Countryside Cultures*. London: Routledge.

Siebert, R., Laschewski, L. and Dosch, A. (2008) Knowledge dynamics in valorising local nature, *Sociologia Ruralis*, 48: 223–39.

Silvasti, T. (2003) Bending boundaries of gender labour division on farms, *Sociologia Ruralis*, 43: 154–66.

Skaptadóttir, U.D. and Wojtynska, A. (2008) Labour migrants negotiating places and engagements, in Bærenholdt, J.O. and Granås, B. (eds) *Mobility and Place: Enacting Northern European Peripheries*. Aldershot: Ashgate.

Slater, D. (1974) Colonialism and the spatial structure of underdevelopment: outlines of an alternative approach, with special reference to Tanzania, *Progress in Planning*, 4: 146–59.

Slocum, R. (2008) Thinking race through corporeal feminist theory: divisions and intimacies at Minneapolis Farmers' Market, *Social and Cultural Geography*, 9: 849–69.

Smith, D. (2007) The changing faces of rural populations: "(re) fixing' the gaze' or 'eyes wide shut'? *Journal of Rural Studies*, 23: 275–82.

Smith, D.P. and Phillips, D.A. (2001) Socio-cultural representations of greentrified Pennine rurality, *Journal of Rural Studies*, 17: 457–70.

Smith, E. and Marsden, T. (2004) Exploring the 'limits to growth' in UK organics beyond the statistical image, *Journal of Rural Studies*, 20: 345–58.

Smith, H.A. and Furuseth, O.J. (eds) (2006) *Latinos in the New South*. Burlington, VT: Ashgate.

Smith, M.J. (1989) Changing policy agendas and policy communities: agricultural issues in the 1930s and 1980s, *Public Administration*, 67: 149–65.

Smith, S.J. (1993) Bounding the Borders: claiming space and making place in rural Scotland, *Transactions of the Institute of British Geographers*, 18: 291–308.

Smith, W. and Montgomery, H. (2003) Revolution or evolution? New Zealand agriculture since 1984, *Geojournal*, 59: 107–18.

Smithers, J., Joseph, A.E. and Armstrong, M. (2005) Across the divide(?): reconciling farm and town views of agriculture – community linkages, *Journal of Rural Studies*, 21: 281–95.

Snowden, F.M. (2007) *The Conquest of Malaria: Italy, 1900–1962*. New Haven: Yale University Press.

Southgate, D., Graham, D.H. and Tweenten, L. (2007) *The World Food Economy*. Malden, MA: Blackwell.

Squire, S.J. (1992) Ways of seeing, ways of being: literature, place and tourism in L.M. Montgomery's Prince Edward Island, in Simpson-Housley, P. and Norcliffe, G. (eds) *A Few Acres of Snow: Literary and Artistic Images of Canada*. Toronto: Dundurn Press.

Standlea, D.M. (2006) *Oil, Globalization and the War for the Arctic Refuge*. Albany, NY: State University of New York Press.

Stedman, R.C. (2006) Understanding place attachment among second home owners, *American Behavioral Scientist*, 50: 187–205.

Steger, M.B. (2003) *Globalization: A very short introduction*. Oxford: Oxford University Press.

Stock, C.M. (1996) *Rural Radicals: Righteous rage in the American grain*. London: Cornell University Press.

Stockdale, A. (2004) Rural out-migration: community consequences and individual migrant experiences, *Sociologia Ruralis*, 44: 167–94.

Stockdale, A. (2006) Migration: pre-requisite for rural economic development? *Journal of Rural Studies*, 22: 354–66.

Storey, D. (2004) A sense of place: rural development, tourism and place promotion in the Republic of Ireland, in Holloway, L. and Kneafsey, M. (eds) *Geographies of Rural Cultures and Societies*. Aldershot: Ashgate.

Storey, D. (2010) Partnerships, people and place: lauding the local in rural development, in Halseth, G., Markey, S. and Bruce, D. (eds) *The Next Rural Economies: Constructing rural place in global economies*. Wallingford: CABI.

Svendsen, G. (2004) The right to development: construction of a non-agriculturalist discourse of rurality in Denmark, *Journal of Rural Studies*, 20: 79–94.

Taylor, P.J. (1989) The error of developmentalism in human geography, in Gregory, D. and Walford, R. (eds) *Horizons in Human Geography*. London: Macmillan.

Thrift, N. (1989) Images of social change, in Hamnett, C., McDowell, L. and Sarre, P. (eds) *The Changing Social Structure*. London: Sage.

Thrift, N. (2007) *Non-representational Theory: Space, Power, Affect*. London: Routledge.

Tönnies, F. (1963) *Community and Society*. New York: Harper and Row.

Torres, R.M., Popke, J. and Hapke, H.M. (2006) The South's Silent Bargain: Rural restructuring, Latino labor and the ambiguities of migrant experience, in Smith, H.A. and Furuseth, O.J. (eds) *Latinos in the New South*. Burlington, VT: Ashgate.

Tovey, H. (1997) Food, environmentalism and rural sociology: on the organic farming movement in Ireland, *Sociologia Ruralis*, 37: 21–37.

Tregear, A. (2003) From Stilton to Vimto: Using food history to re-think typical products in rural development, *Sociologia Ruralis*, 43: 91–107.

Tremlett, G. (2005) Spain's greenhouse effect: the shimmering sea of polythene consuming the land, *The Guardian*, 21 September.

Trubek, A.B. and Bowen, S. (2008) Creating the taste of place in the United States: can we learn from the French? *GeoJournal*, 73: 23–30.

Turner, F.J. (1920) *The Frontier in American History*. New York: Henry Holt.

Tykkyläinen, M. (2008) The future of the 'boom and bust' landscape in the Russian north, in Rautio, V. and Tykkyläinen, M. (eds) *Russia's Northern Regions on the Edge*. Helsinki: Kikimora Publications.

Tyler, K. (2006) Village people: race, class, nation and the community spirit, in Agyeman, J. and Neal, S. (eds) *The New Countryside? Ethnicity, nation and exclusion in contemporary rural Britain*. Bristol: Policy Press.

Tzanelli, R. (2004) Constructing the 'cinematic tourist': The 'sign industry' of the Lord of the Rings, *Tourist Studies*, 4: 21–42.

Urbain, J. (2002) *Paradis Verts: Désirs de Campagne et Passions Résidentielles*. Paris: Payot.

Urry, J. (1990) *The Tourist Gaze*. London: Sage.

Urry, J. (2003) *Global Complexity*. Cambridge: Polity.

Valenčius, C.B. (2002) *The Health of the Country*. New York: Basic Books.

Van Dam, F., Heins, S. and Elbersen, B.S. (2002) Lay discourses of the rural and stated and revealed preferences for rural living: some evidence of the existence of a rural idyll in the Netherlands, *Journal of Rural Studies*, 18: 461–76.

Vanderbeck, R.M. (2003) Youth, racism and place in the Tony Martin affair, *Antipode*, 35: 363–84.

Vanderbeck, R.M. (2006) Vermont and the imaginative geographies of American whiteness, *Annals of the Association of American Geographers*, 96: 641–59.

Van der Ploeg, J.D. (2008) *The New Peasantries: Struggles for autonomy and sustainability in an era of empire and globalization*. London: Earthscan.

Van der Ploeg, J.D. and Marsden, T. (eds) (2008) *Unfolding Webs: the dynamics of regional rural development*. Assen: Van Gorcum.

Van der Ploeg, J.D., Renting, H., Brunori, G., Knickel, K., Mannion, J., Marsden, T., de Roest, K., Sevilla Guzmán E. and Ventura, F. (2000) Rural development: from practice and policies to theory, *Sociologia Ruralis*, 40: 391–408.

Veeck, G., Che, D. and Veeck, A. (2006) America's changing farmscape: a study of agricultural tourism in Michigan, *Professional Geographer*, 58: 235–48.

Velayutham, S. and Wise, A. (2005) Moral economies of a translocal village: obligation and shame among South Indian transnational migrants, *Global Networks*, 5: 27–47.

Vinet, F. (2008) Geographical analysis of damage due to flash floods in southern France: the cases of 12–13 November 1999 and 8–9 September 2002, *Applied Geography*, 28: 323–36.

Vitek, W. (1996) 'Rediscovering the landscape', in Vitek, W. and Jackson, W. (eds) *Rooted in the Land*. New Haven: Yale University Press.

Vitek, W. and Jackson, W. (eds) (1996) *Rooted in the Land*. New Haven: Yale University Press.

Waitt, G. and Cook, L. (2007) Leaving nothing but ripples on the water: performing ecotourism natures, *Social and Cultural Geography*, 8: 535–50.

Waitt, G. and Lane, R. (2007) Four-wheel drivescapes: Embodied understandings of the Kimberley, *Journal of Rural Studies*, 23: 156–69.

Walford, N. (2003) Productivism is allegedly dead, long live productivism: Evidence of continued productivist attitudes and decision-making in South-East England, *Journal of Rural Studies*, 19: 491–502.

Walford, N. (2007) Geographical and geodemographic connections between different types of small area and the origins and destinations of migrants to Mid Wales, *Journal of Rural Studies*, 23: 318–31.

Walker, P. and Fortmann, L. (2003) Whose landscape? A political ecology of the 'exurban' Sierra, *Cultural Geographies*, 10: 469–91.

Walker, R.A. (2001) California's golden road to riches: natural resources and regional capitalism, 1848–1940, *Annals of the Association of American Geographers*, 91: 167–99.

Walker, R.A. (2005) *The Conquest of Bread*. New York: The New Press.

Warner, M.E. (2006) Market-based governance and the challenge for rural governments: US trends, *Social Policy and Administration*, 40: 612–31.

Waters, T. (2007) *The Persistence of Subsistence Agriculture*. Lanham, MD: Lexington Books.

Weis, T. (2007) *The Global Food Economy*. London: Zed Books.

Whatmore, S. (2002) *Hybrid Geographies*. London: Sage.

Whitehead, M., Jones, R. and Jones, M. (2008) *The Nature of the State: Excavating the political ecologies of the modern state*. Oxford: Oxford University Press.

Whittle, J. (2000) *The Development of Agrarian Capitalism: Land and labour in Norfolk, 1440–1580*. Oxford: Oxford University Press.

Wiborg, A. (2004) Place, nature and migration: students' attachments to their rural home places, *Sociologia Ruralis*, 44: 416–32.

Wilkie, R. (2005) Sentient commodities and productive paradoxes: the ambiguous nature of human-livestock relations in Northeast Scotland, *Journal of Rural Studies*, 21: 213–30.

Williams G. (2008) *Struggles for an Alternative Globalization: An ethnography of counterpower in southern France*. Aldershot: Ashgate.

Williams, R. (1973) *The Country and the City*. New York: Oxford University Press.

Wilson, A. (1992) *The Culture of Nature*. Cambridge, MA: Blackwell.

Wilson, G.A. (2001) From productivism to post-productivism ... and back again? Exploring the (un)changed natural and mental landscapes of European agriculture, *Transactions of the Institute of British Geographers*, 26: 77–102.

Wilson, G.A. (2007) *Multifunctional Agriculture: A transition theory perspective*. Wallingford: CAB International.

Wilson, G.A. (2008) From 'weak' to 'strong' multifunctionality: conceptualising farm-level multifunctional transitional pathways, *Journal of Rural Studies*, 24: 367–83.

Winchester, H.P.M. and Rofe, M.W. (2005) Christmas in the 'Valley of Praise': Intersections of rural idyll, heritage and community in Lobethal, South Australia, *Journal of Rural Studies*, 21: 265–79.

Winders, B. (2004) Sliding towards the free market: shifting political coalitions and US agricultural policy, 1945–75, *Rural Sociology*, 69: 467–89.

Winter, M. (1996) *Rural Politics*. London: Routledge.

Winter, M. (2003) Embeddedness, the new food economy and defensive localism, *Journal of Rural Studies*, 19: 23–32.

Wirth, L. (1938) Urbanism as a way of life, *American Journal of Sociology*, 44: 1–24.

Wittman, H. (2009) Reframing agrarian citizenship: Land, life and power in Brazil, *Journal of Rural Studies*, 25: 120–30.

W.K. Kellogg Foundation (2002) *Perceptions of Rural America*. Battle Creek, MI: W.K. Kellogg Foundation.

Wolford, W. (2004) The land is ours now: spatial imaginaries and the struggle for land in Brazil, *Annals of the Association of American Geographers*, 94: 409–24.

Woods, M. (1998a) Researching rural conflicts: hunting, local politics and actor-networks, *Journal of Rural Studies*, 14: 321–40.

Woods, M. (1998b) Mad cows and hounded deer: political representations of animals in the British countryside, *Environment and Planning A*, 30: 1219–34.

Woods, M. (2000) Fantastic Mr Fox? Representing animals in the hunting debate, in Philo, C. and Wilbert, C. (eds) *Animal Spaces, Beastly Places*. London: Routledge.

Woods, M. (2003a) Deconstructing rural protest: the emergence of a new social movement, *Journal of Rural Studies*, 19: 309–25.

Woods, M. (2003b) Conflicting environmental visions of the rural: windfarm development in Mid Wales, *Sociologia Ruralis*, 43: 271–88.

Woods, M. (2005a) *Rural Geography*, London: Sage.

Woods, M. (2005b) *Contesting Rurality: Politics in the British countryside*. Aldershot: Ashgate.

Woods, M. (2007) Engaging the global countryside: globalization, hybridity and the reconstitution of rural place, *Progress in Human Geography*, 31: 485–507.

Woods, M. (2008a) New Labour's countryside, in Woods, M. (ed.) *New Labour's Countryside: British Rural Policy since 1997*. Bristol: Policy Press.

Woods, M. (2008b) Hunting: New Labour success or New Labour failure?, in Woods, M. (ed.) *New Labour's Countryside: British Rural Policy since 1997*, Bristol: Policy Press.

Woods, M. (2008c) Social movements and rural politics, *Journal of Rural Studies*, 24: 129–37.

Woods, M. (2009a) Rural Geography, in Kitchin, R. and Thrift, N. (eds) *International Encylopedia of Human Geography*, vol. 9. Oxford: Elsevier.

Woods, M. (2009b) Exploring the uneven geographies of 'rural geography': commentary on M. Kurtz and V. Craig, 'Constructing rural geographies in publication', *ACME: An International E-Journal for Critical Geographies*, 8: 394–413.

Woods, M. (2010a) The political economies of place in the emergent global countryside: stories from rural Wales, in Halseth, G., Markey, S. and Bruce, D. (eds) *The Next Rural Economies: Constructing rural place in global economies*. Wallingford: CABI.

Woods, M. (2010b) Representing Rural America: the reconstruction of a political space, in Winchell, D. Ramsay, D., Koster, R. and Robinson, G. (eds) *Sustainable Rural Community Change: Geographical perspectives from North America, the British Isles and Australia*. Brandon: Rural Development Institute.

Woods, M. (2010c) The local politics of the global countryside: boosterism, aspirational ruralism and the contested reconstitution of Queenstown, New Zealand, *Geojournal*, advance online publication, DOI: 10.1007/s10708-009-9268-7.

Woods, M., Edwards, B., Anderson, J. and Gardner, G. (2007) Leadership in place: elites, institutions and agency in British rural community governance, in Cheshire, L., Lawrence, G. and Higgins, V. (eds) *Rural Governance: International Perspectives*. London: Routledge.

Woods, M. and Goodwin, M. (2003) Applying the rural, in Cloke, P. (ed.) *Country Visions*. Harlow: Pearson.

Wylie, J. (2005) A single day's walking: narrating self and landscape on the South West Coast Path, *Transactions of the Institute of British Geographers*, 30: 234–47.

Wylie, J. (2007) *Landscape*. London: Routledge.

Xu, W. and Tan, K.C. (2002) Impact of reform and economic restructuring on rural systems in China: a case study of Yuhang, Zhejiang, *Journal of Rural Studies*, 18: 65–82.

Yarwood, R. and Charlton, C. (2009) 'Country life'? Rurality, folk music and 'Show of Hands', *Journal of Rural Studies*, 25: 194–206.

Zaccaro, S. (2009) *Development: Think of the Women Farmers*. IPS News, 19 February. Available http: <http:\\ipsnews.net>

Zezza, A., Carletto, G., Davis, B., Stamoulis, K. and Winters, P. (2009) Rural income generating activities: whatever happened to the institutional vacuum? Evidence from Ghana, Guatemala, Nicaragua and Vietnam, *World Development*, 37: 1297–306.

Zografos, C. and Martinez-Alier, J. (2009) The politics of landscape value: a case study of wind farm conflict in Catalonia, *Environment and Planning A*, 41: 1726–44.

Zukin, S. (1990) Socio-spatial prototypes of a new organization of consumption: the role of real cultural capital, *Sociology*, 24: 37–56.

Index

Routledge

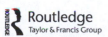

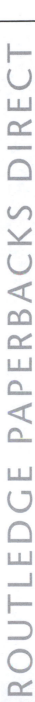